Proceedings of the 2019 Meeting of International Society for Data Science and Analytics

I0032184

New Developments in Data Science and Data Analytics

Zhiyong Zhang
Ke-Hai Yuan
Yong Wen
Jiashan Tang

ISDSA Press · Granger, IN

Copyright © 2020 International Society for Data Science and Analytics
First printing, May, 2020

PUBLISHED BY ISDSA PRESS GRANGER, IN
WWW.ISDSA.ORG

ISBN-13 (pbk): 978-1-946728-03-6
ISBN-13 (E-book): 978-1-946728-04-3

This work is subject to copyright. All rights are reserved by the ISDSA Press and the authors, whether the whole or part of the material is concerned, specifically the rights of translation, reprinting, reuse of illustrations, recitation, broadcasting, reproduction on microfilms or in any other physical way, and transmission or information storage and retrieval, electronic adaptation, computer software, or by similar or dissimilar methodology now known or hereafter developed. Exempted from this legal reservation are brief excerpts in connection with reviews or scholarly analysis or material supplied specifically for the purpose of being entered and executed on a computer system, for exclusive use by the purchaser of the work. Duplication of this publication or parts thereof is permitted only under the provisions of the Copyright Law of the Publisher's location, in its current version, and permission for use must always be obtained from ISDSA Press. Violations are liable to prosecution under the respective Copyright Law.

Trademarked names, logos, and images may appear in this book. Rather than use a trademark symbol with every occurrence of a trademarked name, logo, or image we use the names, logos, and images only in an editorial fashion and to the benefit of the trademark owner, with no intention of infringement of the trademark.

While the advice and information in this book are believed to be true and accurate at the date of publication, neither the authors nor the editors nor the publisher can accept any legal responsibility for any errors or omissions that may be made. The publisher makes no warranty, express or implied, with respect to the material contained herein.

Library of Congress Cataloging-in-Publication Data

Personal name: Zhang, Zhiyong (Professor), author
Main title: New Developments in Data Science and Data Analytics / Zhiyong Zhang, Ke-Hai Yuan, Yong Wen, and Jiashan Tang
Published/Produced: ISDSA Press
LCCN: 2019905671
LCCN Permalink: https://lccn.loc.gov/2019905671

Preface

This conference proceeding represents presentations given at the Annual Meeting of the International Society for Data Science and Analytics (ISDSA) in Nanjing, China, during July 6–8, 2019. The Annual Meeting of ISDSA aims to provide a global forum for researchers and practitioners in the field of data science and data analytics to communicate and showcase their latest research. The meeting was designed to have a single plenary session with invited talks to maximize intellectual exchanges.

The 2019 Meeting had more than 50 attendees with 28 presentations. The researchers were from different academic disciplines and industries from multiple countries. They presented applications and methods in the areas of education, environmental research, medical research, political science, psychology, and so on. Two best paper awards were given to Dr. Xin Gu, Associate Professor from East China Normal University, and Dr. Zijun Ke, Assistant Professor from Sun Yet-Sen University. This book collects 11 papers presented in the meeting.

Acknowledgments

We first thank all the participants of the meeting, without whom the meeting would not have been possible and successful. We also thank the organizing committee members who have reviewed the submitted abstracts including Dr. Han Du from University of California, Los Angeles, Dr. Ge Jiang from University of Illinois at Urbana-Champaign, Dr. Haiyan Liu from University of California, Merced, Dr. Hongyun Liu from Beijing Normal University (China), Dr. Laura Lu from University of Georgia, Dr. Xin Tong from University of Virginia, Dr. Qian Zhang, Florida State University, and Drs. Yun Sun and Xuetao Yang from Nanjing University of Posts and Telecommunications (China). Several student volunteers helped with the logistics of the meeting. They are Bingjiang Li, Chengfang Zhu, Raomin Hu, and Lu Peng. Bingjiang Li, Jing Chen, Peng Xu, and Jing Wang also assisted with the publication of the proceeding book. Finally, we want to thank Nanjing University of Posts and Telecommunications and University of Notre Dame for providing financial support to the meeting.

Zhiyong Zhang and Ke-Hai Yuan
University of Notre Dame

Yong Wen and Jiashan Tang
Nanjing University of Posts and Telecommunications

April, 2020

List of Presentations

Below is a full list of presentations, ordered by the time of presentation, at the 2019 Annual Meeting of ISDSA. Full information about each presentation can be found on the meeting website at `https://meeting.isdsa.org`.

Prospect Theory and Stock Returns: A Test for Trading Behavior of Individual VS. Institutional Investors
Xiaoling Zhong, University of Science and Technology of China
Junbo Wang, City University of Hong Kong

A Nonparametric Multivariate Statistical Process Control Chart Based on Change Point Model
Yafei Xu, Beijing AI Lab, Vivo Communication Technology Co. Ltd., China

WeibullR: An R Package for Weibull Analysis for Reliability Engineering
David Silkworth, OpenReliability.org, United States

Evaluating Informative Hypotheses Using the Bayes Factor
Xin Gu, East China Normal University, China

A Slice Inverse Regression Algorithm Based on k-Medoids Clustering
Jiashan Tang, Nanjing University of Posts and Telecommunications, China

Improving Teaching Evaluation using Text Mining
Zhiyong Zhang, University of Notre Dame, United States

Comparing Three MASEM Approaches to Quantifying or Explaining Between-study Heterogeneity in SEM Parameters
Zijun Ke, Sun Yat-Sen University, China

Model Uncertainty in the Comparison of Two Single Dengue Outbreaks
Carlos Rafael Sebrango Rodríguez, Universidad de Sancti Spiritus "José Martí Pérez", Cuba
Lizet Sánchez Valdés, Centro de Inmunología Molecular, Cuba
Ziv Shkedy, Hasselt University, Belgium
Vivian Sistachs Vega, Universidad de La Habana, Cuba

An Evaluation of Statistical Differential Analysis Methods in Single-cell RNA-seq Data
Dongmei Li, University of Rochester, United States

Exploring Spatio-Temporal Patterns of Air Quality Index Data in China
Haokun Tang, Yulin Xie, and Binbin Lu, Wuhan University, China

A Review of Aspect-Based Sentiment Analysis with an Application on Teaching Evaluation
Wen Qu, University of Notre Dame, United States

A Non-Parametric Model to Address Overdispersed Count Response in a Longitudinal Data Setting with Missingness
Hui Zhang, Northwestern University, United States

A General Bayesian Model-Based Imputation Approach for Multilevel Models with Non-linear Effects: A Sequential Approach
Han Du and Craig Enders, University of California, Los Angeles, United States

MCMC Bootstrap Based Approach to Power/Sample Size Evaluation
Oleksandr Mykolayovich Ocheredko, Vinnytsya National Medical University, Ukraine

Advances of Social Network Analysis in Psychological Sciences
Haiyan Liu, University of California, Merced, United States

Correlation Analysis between Tourism Development, Economic Growth and Carbon Emissions: A Comparative Analysis Based on Six Provinces in the Central China
Zhibiao Wang, Yangtze Normal University, China
Peibo Yao, Henan University, China

Mediation Analysis for Complex Surveys with Balanced Repeated Replications
Yujiao Mai, St. Jude Children's Research Hospital, United States

Table of Contents

Latent Growth Curve Models with VAR Residuals for Longitudinal Mediation Analysis

Xiao Liu[0000−0002−4526−3221]

University of Notre Dame, Notre Dame IN 46556, USA
xliu19@nd.edu

Abstract. Mediation analysis using longitudinal data has become increasingly popular. To perform longitudinal mediation analysis, different models have been proposed, such as the latent growth curve mediation model (LGCM) and the cross-lagged panel mediation model (CLPM). In the current study, we proposed an alternative longitudinal mediation model (referred to as LGCM-CLRM), where a system of latent growth curve models are used to describe the deterministic inherent trajectories of each individual and a vector autoregressive model is used to describe the within-individual stochastic deviations from the latent trajectory. Compared to the existing longitudinal mediation models, the proposed model allows the mediation effects in both level-1 and level-2 models, and thus could disentangle different types of mediation effects. The proposed model can be estimated in the multilevel structural equation modeling framework. Simulation studies were performed to evaluate the estimation quality. We also provided a real data example for illustration.

Keywords: Longitudinal mediation analysis · Growth curve modeling · Vector autoregressive residuals.

DOI: 10.35566/isdsa2019c1

1   Introduction

Mediation analysis (Baron & Kenny, 1986) is widely used by researchers in various fields. It models the phenomenon that the effect of the independent variable X on the dependent variable Y is transmitted through a third variable M. This kind of effect, which is often termed as mediation or indirect effect in psychological and behavioral sciences, could imply a causal hypothesis that the independent variable X causes the mediating variable M, which in turn causes the dependent variable Y (Sobel & Lindquist, 2014). Such hypotheses that articulate measurable processes intervening between the independent and dependent variables are fundamental to theories in many substantive studies.

Both cross-sectional and longitudinal designs have been used in substantive investigations of mediation. Although cross-sectional mediation studies may be convenient in data collection, cross-sectional mediation analyses often fail to consider the role of time. However, mediational processes are inherently causal, implying temporal successions of the variables involved. Without appropriately accounting for the temporal conditions, the conclusion from an empirical mediation analysis can be misleading.

In contrast, mediation analyses using longitudinal data allow the opportunity of explicitly taking the temporal order assumption of mediation into consideration, and thus could provide more valid causal inference. Furthermore, for longitudinal data from multiple individuals, mediation models could have multiple indirect effects and even different types of indirect effects.

1.1   Different Frameworks for Longitudinal Mediation Analysis

To perform mediation analysis using longitudinal data, different longitudinal mediation models have been proposed. Among different longitudinal mediation models, three are most widely used in psychology and behavioral sciences: the latent growth curve mediation model LGCM; e.g., von Soest and Hagtvet (2011), the cross-lagged panel mediation model (CLPM; e.g., Maxwell and Cole (2007)), and the latent difference score mediation model (LDSM; e.g., McArdle and Hamagami (2001)). Generally, the latent growth models are particularly suited to mediation analyses where individual trajectories exhibit a relatively large amount of intra-individual change. In contrast, cross-lagged panel mediation models and latent difference score mediation models are suggested be favored when the change from one time point to the next is to be assessed (Selig & Preacher, 2009). The choice of longitudinal mediation models is generally based on the theory of change, the span of the study, and the lag between adjacent measurement occasions.

2   An Alternative Longitudinal Mediation Model

In the current study, we considered an alternative longitudinal mediation model that can be viewed as a combination of LGCM and CLPM. We refer to the model as LGCM-CLRM ("R" represents within-individual "residuals") in the rest of the article. Suppose the independent variable X, the mediator M, and the outcome variable Y are longitudinal processes and multivariate time series data are collected for X, M, Y from N individuals: $\{X_{i,t}, M_{i,t}, Y_{i,t}\}_{t=1}^{T_i}, i = 1, ..., N$. We consider separating each observed score $\{X_{i,t}, M_{i,t}, Y_{i,t}\}$ into two parts: a deterministic part modeled as multivariate latent growth curves and a stochastic part modeled as a stationary multivariate time series. Specifically, the longitudinal mediation model that we considered is a multivariate linear latent growth curve model where the (multivariate) within-individual residuals follow a vector autoregressive model of order 1 (VAR(1)). Mediation effects are allowed to exist in both the multivariate latent growth curves and the multivariate within-individual residuals, and thus level-2 and level-1 mediation effects can be disaggregated. We considered a VAR model of order 1 to model mediation effects in the within-individual residuals, because in the substantive applications of the CLPM model, the autoregressive effects and the cross-lagged effects are often of lag 1. For the t-th measurement of individual i ($i = 1, \ldots, N$), the level-1 model is expressed as:

$$\begin{pmatrix} X_{i,t} \\ M_{i,t} \\ Y_{i,t} \end{pmatrix} = \begin{pmatrix} I_i^X \\ I_i^M \\ I_i^Y \end{pmatrix} + \begin{pmatrix} S_i^X \\ S_i^M \\ S_i^Y \end{pmatrix} Time_{i,t} + \mathbf{u}_{i,t}, \quad t = 1, \ldots, T_i$$

where I_i^X, I_i^M, I_i^Y denote the latent intercepts and S_i^X, S_i^M, S_i^Y the latent slopes.

Before specifying the level-2 model, we note that the specification of $Time_{i,t}$ could influence the interpretation of parameters in the latent growth curve models. In the current study, we focus on the condition $T_i = T$ for all individuals $i = 1, \ldots, N$ (i.e., there is no between-individual difference of the level-1 predictor $Time_{i,t}$). We used $Time_{i,t} = t - 1$, so that intercept can be defined to provide only information about the baseline level. Then, the level-2 mediation model that we considered is:

$$\begin{aligned} I_i^X &= I_0^X + v_i^{IX}, & S_i^X &= S_0^X + v_i^{SX}, \\ I_i^M &= I_0^M + v_i^{IM}, & S_i^M &= S_0^M + a^{SI} I_i^X + v_i^{SM}, \\ I_i^Y &= I_0^Y + v_i^{IY}, & S_i^Y &= S_0^Y + c^{SI} I_i^X + b^{SS} S_i^M + v_i^{SY}. \end{aligned}$$

In the level-2 model, the mediation pathway is $I_i^X \to S_i^M \to S_i^Y$, where a^{SI} represents the effects of initial levels of X on the changes in M, b^{SS} represents the effects of changes in M on the changes in Y, holding the initial levels of X constant, and c^{SI} represents the effects of initial levels of X on the changes in Y, holding the changes in M constant. We let I_0^X, I_0^M, I_0^Y denote the population mean of the latent intercepts, S_0^X, S_0^M, S_0^Y the population mean of the latent slopes, and $v_i^{IX}, v_i^{SX}, v_i^{IM}, v_i^{SM}, v_i^{IY}, v_i^{SY}$ the level-2 residuals of the latent intercepts and slopes.

For the level-2 residual terms, we assume that the random vector $\mathbf{v}_i = (v_i^{IX}, v_i^{SX}, v_i^{IM}, v_i^{SM}, v_i^{IY}, v_i^{SY})'$ is independent of the level-1 residuals $\mathbf{u}_{i,t}$, and independently follows a multivariate normal distribution with mean zero

$$E\left(\mathbf{v}_i\right) = \mathbf{0}$$

and the following covariance matrix (the lower-triangular part is the same as the upper-triangular part):

$$\text{Cov}\left(\mathbf{v}_i\right) = \begin{pmatrix} \sigma_{II}^X & \sigma_{IS}^X & 0 & 0 & 0 & 0 \\ & \sigma_{SS}^X & 0 & 0 & 0 & 0 \\ & & \sigma_{II}^M & \sigma_{IS}^M & 0 & 0 \\ & & & \sigma_{SS}^M & 0 & 0 \\ & & & & \sigma_{II}^Y & \sigma_{IS}^Y \\ & & & & & \sigma_{SS}^Y \end{pmatrix}.$$

Therefore, we assume the specified mean structure is correct and there are no omitted confounders in the level-2 model.

Follow from the above model specifications and assumptions, the model implied mean structure for individual i is:

$$
\mathrm{E}\begin{pmatrix} \mathbf{X}_i \\ \mathbf{M}_i \\ \mathbf{Y}_i \end{pmatrix} = \mathbf{Z}_i \boldsymbol{\beta},
$$

where

$$
\mathbf{Z}_i = \begin{pmatrix}
1 & 0 & & & & \\
 & & 1 & 0 & & \\
 & & & & 1 & 0 \\
\vdots & \vdots & \vdots & \vdots & \vdots & \vdots \\
1 & T_i-1 & & & & \\
 & & 1 & T_i-1 & & \\
 & & & & 1 & T_i-1
\end{pmatrix},
$$

and the vector of fixed effects: $\boldsymbol{\beta} = (I_0^X, S_0^X, I_0^M, S_0^M + a^{SI} I_0^X, I_0^Y, S_0^Y + b^{SS}(S_0^M + a^{SI} I_0^X) + c^{SI} I_0^X)$. The model implied between-individual covariance matrix is $\mathbf{Z}_i \mathbf{D} \mathbf{Z}_i'$, where

$$
\mathbf{D} = \begin{pmatrix}
\sigma_{II}^X & \sigma_{IS}^X & 0 & a^{SI}\sigma_{II}^X & 0 & (c^{SI}+a^{SI}b^{SS})\sigma_{II}^X \\
 & \sigma_{SS}^X & 0 & a^{SI}\sigma_{IS}^X & 0 & (c^{SI}+a^{SI}b^{SS})\sigma_{IS}^X \\
 & & \sigma_{II}^M & \sigma_{IS}^M & 0 & b^{SS}\sigma_{IS}^M \\
 & & & (a^{SI})^2\sigma_{II}^X + \sigma_{SS}^M & 0 & a^{SI}(c^{SI}+a^{SI}b^{SS})\sigma_{II}^X + b^{SS}\sigma_{SS}^M \\
 & & & & \sigma_{II}^Y & \sigma_{IS}^Y \\
 & & & & & (c^{SI}+a^{SI}b^{SS})^2\sigma_{II}^X + (b^{SS})^2\sigma_{SS}^M + \sigma_{SS}^Y
\end{pmatrix}.
$$

We assume that, for each individual i, the multivariate within-individual residuals $\{\mathbf{u}_{i,t}\} = \{(u_{i,t}^X, u_{i,t}^M, u_{i,t}^Y)'\}, t = 1, \ldots, T_i$ are stationary after removing the latent trend component represented by the latent linear growth curves. A VAR(1) model was considered for the multivariate within-individual residuals

$$
\mathbf{u}_{i,1} = \Phi_i^{[1]}\mathbf{u}_{i,0} + \epsilon_{i,1}, t = 1
$$
$$
\mathbf{u}_{i,t} = \Phi_i \mathbf{u}_{i,t-1} + \epsilon_{i,t}, t = 2, \ldots, T_i
$$

In the above VAR(1) model, to allow different assumptions for the deviation processes prior to the first measurement occasion, we let $\mathbf{u}_{i,0}$ be the state vector before the first measurement occasion (where stationarity may or may not be reached). The sequence of random shock vectors are assume to be a white noise process: $\{\epsilon_{i,t}\} \sim WN(0, \Psi_i)$ with $\Psi_1 = \ldots = \Psi_N = \Psi$ the same for different individuals. The autoregressive coefficient matrix Φ_i are assumed to be the same for different individuals:

$$
\Phi_1 = \Phi_2 = \ldots = \Phi_N = \Phi = \begin{pmatrix} \phi^X & & \\ a^{UU} & \phi^M & \\ c^{UU} & b^{UU} & \phi^Y \end{pmatrix}.
$$

With the above specifications, if assuming the sequence of within-individual residuals $\{\mathbf{u}_{i,t}\} = \{(u_{i,t}^X, u_{i,t}^M, u_{i,t}^Y)'\}$ is stationary before the first measurement occasion $t = 1$ (i.e., $\phi_i^{[1]} = \Phi_i$), the VAR(1) model for the within-individual residuals can be written more explicitly as:

$$u_{i,t}^X = \phi^X u_{i,t-1}^X + \epsilon_{i,t}^X$$
$$u_{i,t+1}^M = \phi^M u_{i,t}^M + a^{UU} u_{i,t}^X + \epsilon_{i,t+1}^M$$
$$u_{i,t+2}^Y = \phi^Y u_{i,t+1}^Y + + b^{UU} u_{i,t+1}^M + c^{UU} u_{i,t+1}^X + \epsilon_{i,t+2}^Y,$$

where $u_{i,t}^X$, $u_{i,t}^M$, and $u_{i,t}^Y$ denote the stochastic deviations from the inherent change trajectories of X, M and Y at time t, respectively. That is, the level-1 mediation effects are modeled as the cross-lagged effects in the VAR(1) weight matrix (i.e., Granger-type causality), with a^{UU} being the cross-lagged association from $u_{i,t}^X$ to $u_{i,t+1}^M$, b^{UU} the cross-lagged association from $u_{i,t+1}^M$ to $u_{i,t+2}^Y$, and c^{UU} the cross-lagged association from $u_{i,t+1}^X$ to $u_{i,t+2}^Y$.

Following du Toit and Browne (2007), the covariance matrix Ω_i of the ($3T_i \times 1$) vector of within-individual residuals $\{\mathbf{u}_{i,t}\} = \{(u_{i,t}^X, u_{i,t}^M, u_{i,t}^Y)'\}, t = 1, \ldots, T_i$ can be written as:

$$\Omega_i = T_{-\Phi}^{-1} \left[\begin{pmatrix} \mathbf{I}_3 \\ \mathbf{0} \\ \vdots \\ \mathbf{0} \end{pmatrix} \Theta \left(\mathbf{I}_3\ \mathbf{0}\ \cdots\ \mathbf{0} \right) + \mathbf{I}_{T_i} \otimes \Psi \right] T_{-\Phi}^{-1'},$$

where

$$T_{-\Phi} = \begin{pmatrix} \mathbf{I}_3 & & & \\ -\Phi & \mathbf{I}_3 & & \\ & \ddots & \ddots & \\ & & -\Phi & \mathbf{I}_3 \end{pmatrix}_{3T_i \times 3T_i},$$

$\Theta = Var(\Phi_i^{[1]} \mathbf{u}_{i,0})$ is the covariance matrix of the initial state vector ($\otimes$ represents Kronecker product). If assuming the sequence $\{\mathbf{u}_{i,t}\}$ is stationary before the first observation, variance of the initial state vector is $\text{vec}(\Theta) = (\mathbf{I}_{3^2} - \Phi \otimes \Phi)^{-1}\text{vec}(\Phi\Psi\Phi')$, and hence

$$\text{vec}(\Omega_i) = (\mathbf{I}_{3T_i} - T_{-\Phi} \otimes T_{-\Phi})^{-1}\text{vec}(\mathbf{I}_{T_i} \otimes \Psi).$$

Therefore, the model implied covariance for individual i is

$$\text{COV} \begin{pmatrix} \mathbf{X}_i \\ \mathbf{M}_i \\ \mathbf{Y}_i \end{pmatrix} := V_i = Z_i D Z_i' + \Omega_i,$$

or using the vec operator: $\text{vec}(V_i) = (Z_i \otimes Z_i)\text{vec}(D) + \text{vec}(\Omega_i) = (Z_i \otimes Z_i)\text{vec}(D) + (\mathbf{I}_{3T_i} - T_{-\Phi} \otimes T_{-\Phi})^{-1}\text{vec}(\mathbf{I}_{T_i} \otimes \Psi)$.

The LGCM-CLRM model we considered can be represented in Figure 1.

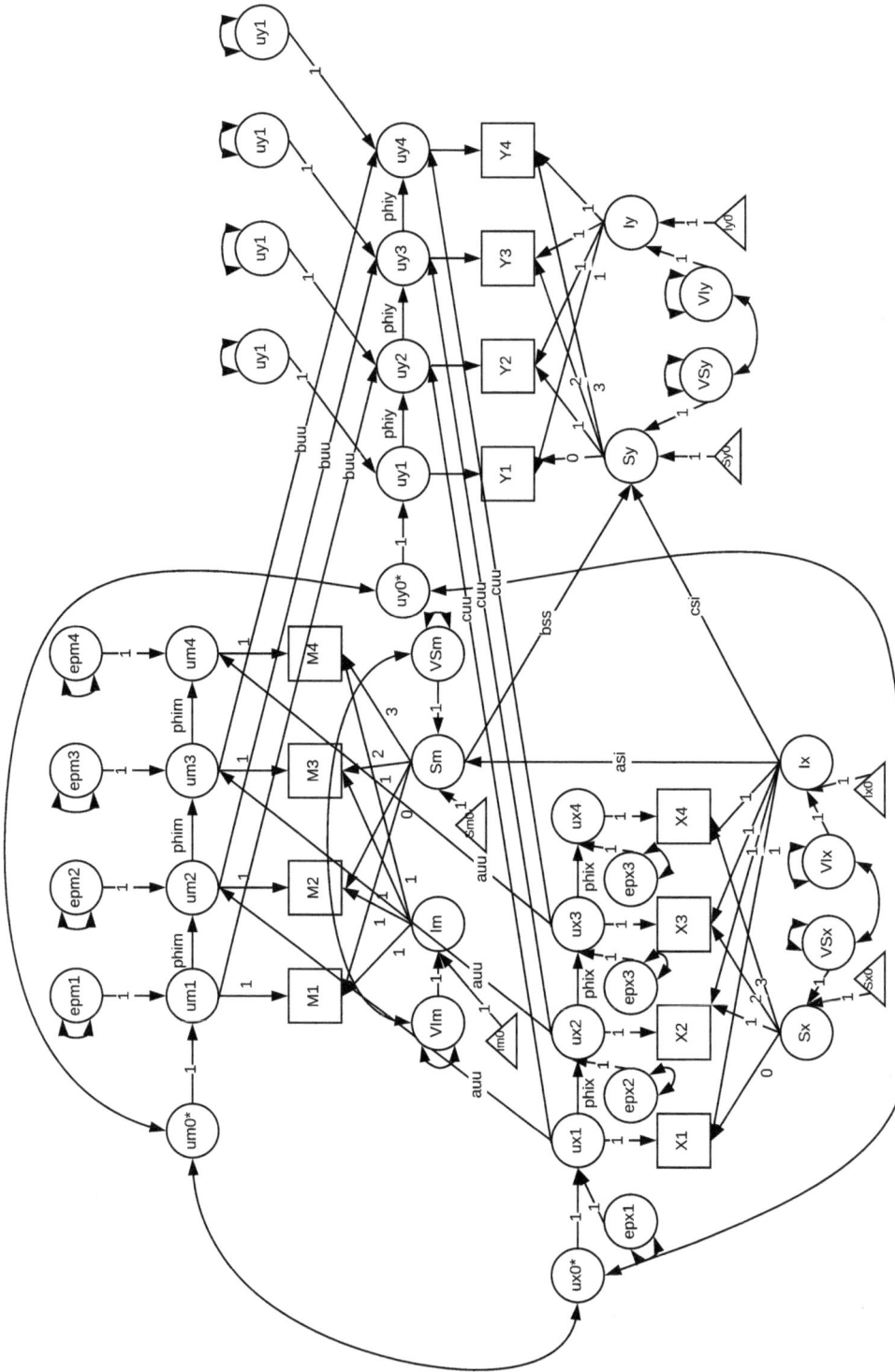

2.1   Equivalent models

There are different ways to add a covariate to a VAR model (Asparouhov, Hamaker, & Muthén, 2018). An equivalent model (i.e., has equivalent fit) of the LGCM-CLRM model we specified above can be written as follows.

For the t-th measurement of individual i $(i = 1, \ldots, N)$, the level-1 models are:

$$\begin{pmatrix} X_{i,t} \\ M_{i,t} \\ Y_{i,t} \end{pmatrix} = \begin{pmatrix} I_i^{X*} \\ I_i^{M*} \\ I_i^{Y*} \end{pmatrix} + \begin{pmatrix} S_i^{X*} \\ S_i^{M*} \\ S_i^{Y*} \end{pmatrix} Time_{i,t} + \Phi \begin{pmatrix} X_{i,t-1} \\ M_{i,t-1} \\ Y_{i,t-1} \end{pmatrix} + \epsilon_{i,t}, \quad t = 1, \ldots, T_i,$$

where the shock variable $\epsilon_{i,t}$ remains the same as the LGCM-CLRM model.

As discussed in Asparouhov et al. (2018), relationships between the random effects in the two equivalent models are

$$\begin{pmatrix} S_i^{X*} \\ S_i^{M*} \\ S_i^{Y*} \end{pmatrix} = (\mathbf{I} - \Phi) \begin{pmatrix} S_i^X \\ S_i^M \\ S_i^Y \end{pmatrix},$$

and

$$\begin{pmatrix} I_i^{X*} \\ I_i^{M*} \\ I_i^{Y*} \end{pmatrix} = (\mathbf{I} - \Phi) \begin{pmatrix} I_i^X \\ I_i^M \\ I_i^Y \end{pmatrix} + \Phi \begin{pmatrix} S_i^X \\ S_i^M \\ S_i^Y \end{pmatrix}.$$

2.2   Model Estimation

The model we considered can be estimated in the multilevel structural equation modeling (ML-SEM) framework (Preacher, Zyphur, & Zhang, 2010). ML-SEM for multilevel mediation is noted for its flexibility. However, in practice, many studies lack the sample size needed to obtain satisfactory estimation results from ML-SEM. Alternatively, the considered model can be estimated in the multivariate linear mixed effects model (MLME) framework. Although MLME and ML-SEM fit the model equivalently, it has been found that MLME has better small sample performance. However, available software packages (e.g., SAS PROC MIXED, R package "lme4" and "nlme") are not directly applicable to estimate MLME models with complex random-effects and residual covariance structures.

When the AR weight matrix and the shock variable covariance matrix are allowed to be random, Bayesian methods can be used to estimate the model. In Mplus 8.2, the model we considered can be estimated as a (residual) dynamic structural equation model (RDSEM) (Asparouhov et al., 2018) using Gibbs Sampling or Metropolis–Hastings algorithm to obtain the posterior distribution.

3   Simulation Studies

To examine the estimation quality of the proposed model, simulation studies were conducted. Three factors were varied in the simulations: (1) complexity of the model, for which we considered free versus stationary initial state vector, (2) number of subjects, where $N = 200, 400$ were considered, and (3) number of time points per subject, where $T = 10, 20$ were used. The following evaluation criteria were used:

- Relative bias $= \frac{\text{Average estimate - True value}}{\text{True value}}$
- $\text{SE}_{estimate}/\text{SE}_{empirical} = \frac{\text{Average standard error estimate}}{\text{Standard deviation of the estimates from the replications}}$
- MSE $= \frac{\sum_{i=1}^{\#reps}(\text{Estimate}_i - \text{True value})^2}{\#reps}$
- Coverage $= \frac{\#\ 95\%\ \text{CI covering true value}}{\#reps}$
- Power $= \frac{\#\ p-value<0.05}{\#reps}$.

In the data generation model, the covariance matrix of the shock variable $\epsilon_{i,t}$ is diagonal

$$\Psi = \begin{pmatrix} \sigma_{\epsilon X}^2 & & \\ & \sigma_{\epsilon M}^2 & \\ & & \sigma_{\epsilon Y}^2 \end{pmatrix},$$

and the state vector $\{\mathbf{u}_{i,t}\} = \{(u_{i,t}^X, u_{i,t}^M, u_{i,t}^Y)'\}$ is stationary before the first measurement occasion. We estimated the model as a ML-SEM model in Mplus 8.2, where equality constraints were imposed such that (1) the times series of the within-individual residuals $\{\mathbf{u}_{i,t}\} = \{(u_{i,t}^X, u_{i,t}^M, u_{i,t}^Y)'\}$ is stationary and (2) covariances of the shock variable are fixed at zero. When assuming the within-individual residuals to be stationary before the first measurement occasion, additional equality constraints were imposed such that elements in the covariance matrix of the initial state vector are functions of VAR(1) model parameters computed based on the stationarity assumption. Results of the simulation study are in the Appendix. Briefly, in the ML-SEM framework, the estimation quality of the proposed LGCM-CLRM model is satisfactory under the simulation conditions we considered.

4   An Empirical Example

The current example was from Empirical Example 2 in O'Laughlin, Martin, and Ferrer (2018). Mediation analysis was performed to investigate whether the effect of high school students' perceived competence in physical education activities (X) on their positive affect of physical education activities (Y) was mediated by their motivation for achievements in physical education activities (M). Two hundred and sixty-one high school students who took a physical education class participated in this study. A total of four waves of data were collected over the course of one school semester. The means, standard deviations, number of observed scores, and intercorrelations for X, M, and Y at each wave of measurement were presented in Table 4 of O'Laughlin et al. (2018).

We fit the LGCM-CLRM model to the sample means and the sample covariance matrix provided in O'Laughlin et al. (2018). For the empirical example, we estimated the model as a ML-SEM model in Mplus 8.2, where equality constraints were imposed such that the observed times series of the within-individual residuals are stationary. For the covariance matrix of the shock vector, we first constrained it to be diagonal and then allowed all the elements of it to be freely estimated. For the covariance matrix of the initial state vector, we first allowed it to be freely estimated, and then constrained the elements of it to be functions of the VAR(1) model parameters based on the stationarity assumption, and also considered it to be a null matrix. Thus, six versions of the proposed LGCM-CLRM model with varied complexities were fitted. However, when we allowed the covariance matrix of the shock vector to be freely estimated and specified the covariance matrix of the initial state vector to be a null matrix, the model estimation did not get convergent results. Under the other five versions of the LGCM-CLRM model, we obtained convergent model estimation results, and presented the parameter estimates and model fit indices in Table 1.

Table 1. Results of six versions of the proposed LGCM-CLRM model for the empirical example

| Shock variable correlations | Zero correlations | | | Freely estimated | |
| Initial state vector | Not sure | Stationarity | Zero | Not sure | Stationarity |
Estimate					
a^{SI}	**1.003**(0.015)	**0.425**(0)	**0.295**(0)	0.019(0.189)	0.044(0.176)
b^{SS}	-0.041(0.356)	**-0.184**(0.014)	-1.452(0.089)	-0.027(0.519)	-0.117(0.3)
c^{SI}	**1.633**(0.016)	**0.681**(0)	**0.892**(0.001)	0.003(0.727)	0.054(0.262)
a^{UU}	**-0.107**(0)	-0.015(0.688)	-0.008(0.822)	0.025(0.239)	**0.079**(0.02)
b^{UU}	0.004(0.632)	**-0.173**(0.019)	-0.015(0.819)	**0.102**(0.024)	**0.222**(0.002)
c^{UU}	-0.102(0.052)	0.008(0.903)	-0.055(0.248)	**0.14**(0)	**0.306**(0)
ϕ^X	0.404(0)	0.38(0)	0.26(0)	0.706(0)	0.646(0)
ϕ^M	0.952(0)	0.434(0)	0.18(0.018)	0.792(0)	0.616(0)
ϕ^Y	0.526(0)	0.058(0.522)	-0.099(0.284)	-0.023(0.809)	-0.071(0.455)
Fit Indices					
Chi-Square	210.458	372.399	330.811	118.058	159.426
df	57	63	63	54	60
RMSEA	0.116	0.157	0.146	0.077	0.089
RMSEA 90% CI	(0.1,0.133)	(0.142,0.173)	(0.131,0.162)	(0.058,0.096)	(0.072,0.106)
CFI	0.857	0.711	0.75	0.94	0.912
TLI	0.834	0.697	0.738	0.927	0.903

The above model estimation results showed that, different assumptions for the shock variable covariance matrix and the influence of the initial state vector led to different model fits and different inferences of the level-1 and the level-2 mediation effects.

In terms of model fit, when assuming the shock variables to be uncorrelated, the model did not fit the data well based on the fit indices (RMSEA, CFI and TLI), for all the three specifications of the influence of the initial state vector. When allowing the covariances of the shock variables to be freely estimated, the model fit improved for both the freely estimated and the stationary initial state vector covariance matrix.

In terms of the inference of mediation effects, when assuming the shock variables were uncorrelated and covariance matrix of the initial state vector satisfied stationary assumption, the level-2 mediation effect was significant (both a^{SI} and b^{SS} were significant) but the level-1 mediation effect was not (a^{UU} was not significant, though b^{UU} was significant). That is, the effect of the initial state of a high school student's perceived competence in physical activities on the linear growth rate of the high school student's enjoyment in physical activities was mediated by the rate of change of the motivation. When allowing the shock variables to be correlated and constraining covariance matrix of the initial state vector to satisfy stationary assumption, the level-2 mediation effect was not significant (neither a^{SI} nor b^{SS} was significant) but the level-1 mediation effect was significant (both a^{UU} and b^{UU} were significant). That is, after removing the latent linear trend of a high school student's perceived competence (X), motivation (M) and enjoyment (Y) of physical activities, a high current level of perceived competence in physical activities was associated with a high level of motivation and a high level of enjoyment in physical activities six weeks (the time lag between two measurement occasions used in the empirical example) later; and a high current level of motivation in physical activities was associated with a high level of enjoyment in physical activities six weeks later. Under the other four model specifications, there were no significant mediation effects in both the level-1 and the level-2 models.

5   Discussion

In the current study, we considered an alternative longitudinal mediation model where latent growth curve models were used to describe the trend components of each individual and a VAR(1) model were used to describe the within-individual residuals. Compared to the existing longitudinal mediation models, the proposed model allows mediation effects in both level-1 and level-2 models, and thus could disentangle different types of mediation effects. However, the model considered in the current study is also limited in the following ways.

First, we assume that the VAR(1) model parameters are the same across individuals. However, the inter-individual differences may exist in the AR weight matrix and the shock variable covariance matrix. The estimation and inference of the LGCM-GLRM model with random VAR model parameters are future directions worth investigation.

Second, measurement error is neglected in the current study. The specified VAR(1) model incorporated dynamic errors, but not measurement error. Schuurman and Hamaker (2019) discussed the concepts of reliability and measurement error in the context of dynamic (VAR(1)) models, and the consequences of disregarding measurement error variance in the data. State-space model can be a useful tool for accommodating measurement error in the latent VAR processes.

Third, parameter estimates of the VAR(1) model of the level-1 residuals could provide insights into potential model misspecification, for instance, a misspecified conditional mean structure (Peruggia et al., 2007). In particular, for the model we considered, S_i^X, I_i^M, or I_i^Y could be potential confounders of the between-individual LGCM. Furthermore, besides the latent factors that are constrained to be uncorrelated, unmeasured confounders may also exist.

Fourth, in the current study, a VAR(1) model is specified for the stochastic process of the within-individual residuals. However, other models, for example, a general $VARMA(p,q)$, can be used to describe the stochastic residuals. As argued in Selig and Preacher (2009), the choice of longitudinal mediation model should be informed by the theory of change.

Fifth, for highly unbalanced longitudinal data with unequal time intervals and missing data, continuous time modeling can be considered for performing longitudinal mediation analysis.

Sixth, with the emergence of SEM studies, graphical model has becoming increasingly popular for exploratory analysis of intensive longitudinal data. It allows sparse estimation of contemporaneous (shock covariance), temporal (AR weights), and between-subjects networks structures (random effects covariance).

References

Asparouhov, T., Hamaker, E. L., & Muthén, B. (2018). Dynamic structural equation models. *Structural Equation Modeling: A Multidisciplinary Journal*, *25*(3), 359–388. doi: https://doi.org/10.1080/10705511.2017.1406803

Baron, R. M., & Kenny, D. A. (1986). The moderator–mediator variable distinction in social psychological research: Conceptual, strategic, and statistical considerations. *Journal of Personality and Social Psychology*, *51*(6), 1173. doi: https://doi.org/10.1037/0022-3514.51.6.1173

du Toit, S. H. C., & Browne, M. W. (2007). Structural equation modeling of multivariate time series. *Multivariate Behavioral Research*, *42*(1), 67-101. (PMID: 26821077) doi: https://doi.org/10.1080/00273170701340953

Maxwell, S. E., & Cole, D. A. (2007). Bias in cross-sectional analyses of longitudinal mediation. *Psychological Methods*, *12*(1), 23–44. doi: https://doi.org/10.1037/1082-989x.12.1.23

McArdle, J. J., & Hamagami, F. (2001). Latent difference score structural models for linear dynamic analyses with incomplete longitudinal data. doi: https://doi.org/10.1037/10409-005

O'Laughlin, K. D., Martin, M. J., & Ferrer, E. (2018). Cross-sectional analysis of longitudinal mediation processes. *Multivariate behavioral research*, *53*(3), 375–402. doi: https://doi.org/10.1080/00273171.2018.1454822

Peruggia, M., et al. (2007). Bayesian model diagnostics based on artificial autoregressive errors. *Bayesian Analysis*, *2*(4), 817–841. doi: https://doi.org/10.1214/07-ba233

Preacher, K. J., Zyphur, M. J., & Zhang, Z. (2010). A general multilevel sem framework for assessing multilevel mediation. *Psychological methods*, *15*(3), 209. doi: https://doi.org/10.1037/a0020141

Schuurman, N. K., & Hamaker, E. L. (2019). Measurement error and person-specific reliability in multilevel autoregressive modeling. *Psychological methods*, *24*(1), 70-91. doi: https://doi.org/10.1037/met0000188

Selig, J. P., & Preacher, K. J. (2009). Mediation models for longitudinal data in developmental research. *Research in Human Development, 6*(2-3), 144–164. doi: https://doi.org/10.1080/15427600902911247

Sobel, M. E., & Lindquist, M. A. (2014, 07). Causal inference for fmri time series data with systematic errors of measurement in a balanced on/off study of social evaluative threat. *Journal of the American Statistical Association, 109*(507), 967–976. Retrieved from https://www.ncbi.nlm.nih.gov/pubmed/25506108 doi: https://doi.org/10.1080/01621459.2014.922886

von Soest, T., & Hagtvet, K. A. (2011). Mediation analysis in a latent growth curve modeling framework. *Structural Equation Modeling, 18*(2), 289–314. doi: https://doi.org/10.1080/10705511.2011.557344

Appendix: Simulation Results

Table 1. Simulation results of the level-2 mediation pathway coefficients (i.e., the fixed effects) when we allow the influence of initial state vector to be freely estimated

	True	*N*	*T*	*Power*	*Coverage*	*RBias*	*MSE*	SE_{est}/SE_{emp}
a^{SI}	0.2	200	10	1	0.96	0.023	0.002	1.088
			20	1	0.96	-0.005	0.001	1.052
		400	10	1	0.96	0.037	0.001	1.054
			20	1	0.96	0.025	0.001	1.055
b^{SS}	0.3	200	10	0.96	0.94	0.014	0.007	0.982
			20	0.96	0.97	-0.036	0.005	1.076
		400	10	1	0.93	0.055	0.004	0.976
			20	1	0.94	0.003	0.003	0.956
c^{SI}	-0.1	200	10	0.55	0.9	0.014	0.003	0.887
			20	0.606	0.939	-0.053	0.002	0.931
		400	10	0.93	0.94	0.03	0.001	1.052
			20	0.92	0.93	0.018	0.001	0.911

Table 2. Simulation results of the level-1 mediation pathway coefficients (i.e., the cross-lagged associations) when we allow the influence of initial state vector to be freely estimated

	True	N	T	Power	Coverage	RBias	MSE	SE_{est}/SE_{emp}
a^{UU}	0.5	200	10	1	0.92	0.004	0.001	0.983
			20	1	0.98	0.005	0	1.095
		400	10	1	0.93	0.002	0	0.996
			20	1	0.95	0.001	0	1.025
b^{UU}	0.3	200	10	1	0.94	-0.003	0.001	0.932
			20	1	0.929	0.002	0	1.037
		400	10	1	0.98	0.001	0	1.045
			20	1	0.95	-0.007	0	0.917
c^{UU}	0.1	200	10	0.98	0.96	0.046	0.001	1.153
			20	1	0.949	0.008	0	0.94
		400	10	1	0.98	0.01	0	1.077
			20	1	0.94	-0.006	0	0.957

Table 3. Simulation results of the level-1 autoregressive coefficients when we allow the influence of initial state vector to be freely estimated

	True	N	T	Power	Coverage	RBias	MSE	SE_{est}/SE_{emp}
ϕ^X	0.7	200	10	1	0.96	0.001	0.002	0.976
			20	1	0.919	-0.004	0	0.984
		400	10	1	0.94	0.004	0.001	0.907
			20	1	0.98	0.001	0	1.1
ϕ^M	0.6	200	10	1	0.93	-0.003	0.001	0.941
			20	1	0.929	-0.001	0	0.929
		400	10	1	0.94	-0.001	0	0.985
			20	1	0.88	0	0	0.877
ϕ^Y	0.5	200	10	1	0.94	-0.007	0.001	0.923
			20	1	0.919	-0.003	0	0.902
		400	10	1	0.94	-0.009	0.001	1.055
			20	1	0.92	-0.002	0	0.866

Table 4. Simulation results of the level-2 mediation pathway coefficients (i.e., the fixed effects) when we assume stationarity before the first occasion

	True	N	T	Power	Coverage	RBias	MSE	SE_{est}/SE_{emp}
a^{SI}	0.2	200	10	0.99	0.94	-0.012	0.002	0.942
			20	1	0.97	0.018	0.001	1.023
		400	10	1	0.969	-0.002	0.001	0.996
			20	1	0.96	-0.011	0.001	1.106
b^{SS}	0.3	200	10	0.95	0.93	-0.033	0.007	0.948
			20	0.94	0.89	-0.027	0.007	0.883
		400	10	1	0.958	0.004	0.003	1.028
			20	1	0.97	-0.012	0.003	1.047
c^{SI}	-0.1	200	10	0.6	0.92	-0.003	0.002	0.931
			20	0.64	0.96	-0.023	0.002	0.956
		400	10	0.906	0.927	0.023	0.001	0.983
			20	0.96	0.97	-0.009	0.001	1.107

Table 5. Simulation results of the level-1 mediation pathway coefficients (i.e., the cross-lagged associations) when we assume stationarity before the first occasion

	True	N	T	Power	Coverage	RBias	MSE	SE_{est}/SE_{emp}
a^{UU}	0.5	200	10	1	0.92	-0.014	0.001	0.931
			20	1	0.98	-0.006	0	1.03
		400	10	1	0.958	0.006	0	1.105
			20	1	0.95	0	0	1.06
b^{UU}	0.3	200	10	1	0.94	0.005	0.001	0.967
			20	1	0.96	-0.002	0	1.031
		400	10	1	0.948	0.004	0	1.017
			20	1	0.92	0.001	0	0.941
c^{UU}	0.1	200	10	0.98	0.95	-0.004	0.001	0.999
			20	1	0.95	-0.002	0	0.966
		400	10	1	0.958	-0.024	0	0.977
			20	1	0.96	-0.009	0	0.95

Table 6. Simulation results of the level-1 autoregressive coefficients when we assume stationarity before the first occasion

	True	N	T	Power	Coverage	RBias	MSE	SE_{est}/SE_{emp}
ϕ^X	0.7	200	10	1	0.94	-0.005	0.001	1.027
			20	1	0.96	-0.001	0	1.062
		400	10	1	0.948	-0.007	0.001	1.047
			20	1	0.92	0	0	0.905
ϕ^M	0.6	200	10	1	0.91	0.01	0.001	0.889
			20	1	0.93	0.001	0	0.899
		400	10	1	0.917	0	0	0.944
			20	1	0.95	0	0	1.046
ϕ^Y	0.5	200	10	1	0.92	-0.002	0.001	0.983
			20	1	0.99	-0.003	0	1.101
		400	10	1	0.948	0.005	0.001	0.986
			20	1	0.93	0.001	0	0.915

A Nonparametric Multivariate Statistical Process Control Chart Based on Change Point Model

Ya Fei Xu[1] and Ostap Okhrin[2]

[1] Eleme Beijing, Alibaba Group, China
xyf252918@alibaba-inc.com
[2] Chair of Econometrics and Statistics, Technische Universität Dresden, Germany

Abstract. This article presents a nonparametric control chart based on the change point model, for multivariate statistical process control (MSPC). The main constituent of the chart is the energy test that focuses on the discrepancy between empirical characteristic functions of two random vectors. This new multivariate control chart highlights in three aspects. Firstly, it is nonparametric, requiring no pre-knowledge of the data generating processes. Secondly, this control chart monitors the whole distribution, not just specific characteristics like mean or covariance. Thirdly, it is designed for online detection (Phase II), which is central for real time surveillance of stream data. Simulation study discusses in-control and out-of-control measures in the context of mean shift and covariance shift. The obtained results are compared with benchmarks and strongly advocate the proposed control chart.

Keywords: Change point model · Energy test · Multivariate statistical process monitoring · Phase II statistical process control.

DOI: 10.35566/isdsa2019c2

1 Introduction

Control chart plays a pivotal role in statistical process monitoring. A natural assumption of the sequence of $d-$dimensional vectors $X_1, \ldots, X_t$, that they are identically independently distributed. The assumption of identical distribution is, however, not always fulfilled, and different sub-sequences of vectors follow different distribution, e.g., the portfolio of stock returns before and after crisis, and characteristics of the product before and after re-calibration of the production machine. In practice, the number of these change points is often unknown, for example, when exactly the crisis started, when the production machine was de-calibrated, etc. Hence, the problem, that the control chart tackles, is to identify these change points, which formally can be considered as a separation of the series $X_1, \ldots, X_t$ into diverse segments, where each adjacent pair of segments follows different distributions.

In the early stage, feature research on statistical process control chart can be referred to seminal papers by Shewhart (1931), Shewhart and Deming (1939),

Page (1954), Roberts (1959). Since multivariate process became useful and common in practical quality engineering Woodall and Montgomery (2014) in recent decades, numerous papers have contributed to forward statistical process control (SPC) in the multivariate context. A part of research is based on parametric assumptions, such as Crosier (1988) for multivariate CUSUM and Lowry, Woodall, Champ, and Rigdon (1992) for multivariate EWMA and Zou and Tsung (2011) with the underlying multivariate Gaussian distribution. Qiu and Hawkins (2001), Qiu and Hawkins (2003), Hawkins and Deng (2010) developed change point models with assumed pre-knowledge in the in-control distribution. Another part of research focusing on online nonparametric multivariate change point models can be found in Zou, Wang, and Tsung (2012), M. Holland and Hawkins (2014) and Zhou, Zi, Geng, and Li (2015). A special accumulation of recent papers on nonparametric control chart can be referred to Chakraborti, Qiu, and Mukherjee (2015). Interested reader can find a comprehensive review of nonparametric control chart in Qiu (2017).

For a proper detection of the changes, different statistical tests with different advantages and disadvantages were used, e.g. Student-t test, Bartlett test and Generalized Likelihood Ratio test (Hawkins, Qiu, & Kang, 2003; Hawkins & Zamba, 2005a, 2005b). The existing-research employs the energy test, which is nonparametric, simple in implementation and has good power (Székely & Rizzo, 2004; Zech & Aslan, 2003; Székely & Rizzo, 2013) investigated the energy statistic and the related test and performed the power analysis for distributional equality. Further, Kim, Marzban, Percival, and Stuetzle (2009) showed the satisfactory performance of the test in the rolling window scheme with fixed window size in detection of change points in image data. Matteson and James (2014) and James and Matteson (2015) employed energy test combined with two different clustering schemes in change point retrospective analysis, e.g., the batch analysis (Phase I).

This paper proposes an *Energy Test Control Chart* (ETCC) that is the nonparametric control chart for online detection of multiple change points in multivariate time series. ETCC gathers three attractive features, which in most other tests are fulfilled separately. Firstly, it is nonparametric without the need of pre-knowledge on the process comparing with traditional parametric control charts. Secondly, this control chart monitors *multivariate* time series which is pervasive in practice, e.g., in financial portfolio management. Last but not least, ETCC controls for more general changes in multivariate time series, e.g., the simultaneous surveillance of mean and covariance. The proposed ETCC does *online* monitoring, which can be applied in many areas using real-time data. To the best of our knowledge, this is the first nonparametric control chart which can simultaneously monitor mean and covariance changes in the multivariate distribution in online fashion.

Methodologically, the ETCC is integrated with the maximum energy divergence based permutation test. The later uses discrepancy between empirical characteristic functions of two random vectors with the empirical distribution of the test statistic obtained using permutation samples. This differs from the commonly used rank test. Afterwards, the sequential detection of change points can be conducted under the algorithm introduced by change point model proposed by Hawkins et al. (2003) to perform online detection.

The simulation study investigates the ETCC in detecting mean and covariance shifts (in multivariate Gaussian, Student-t, Gamma and Laplace distributions). The performance of the ETCC is compared with the benchmark control charts including the spatial rank based EWMA (SREWMA) by Zou et al. (2012), the self-starting multivariate minimal spanning tree (SMMST) based control chart by Zhou et al. (2015) and the nonparametric multivariate change point (NPMVCP) model based control chart by M. Holland and Hawkins (2014). Results indicates a very good performance of the proposed chart.

The paper is structured as follows. Section 2 introduces the methodology of the energy test and the preliminary of change point model in two diverse phases (Phase I and II) providing information on benchmark models. A simulation study and corresponding results are presented in Section 3. Section 4 concludes. Some supplementary materials are provided in appendix.

2   Methodology

2.1   Energy Test

The main constituent of every control chart is the underlying test that is used to control characteristics of mean, variance, and/ or the whole distribution. In the very general set-up with d-dimensional vectors $X \sim F_X$ and $Y \sim F_Y$, we aim at testing $H_0 : F_X = F_Y$ versus $H_0 : F_X \neq F_Y$. It is known that the corresponding characteristic functions ϕ_X and ϕ_Y of X and Y, respectively, are uniquely determined from distribution functions (Serfling, 2009). The usage of the divergence between ϕ_X and ϕ_Y to control for the difference between distributions F_X and F_Y becomes an applicable routine. To test directly for the equivalence of F_X and F_Y fails under the curse of dimensionality and often requires the knowledge of F_X and F_Y. Székely and Rizzo (2005) used an integrated weighted distance between two characteristic functions, and showed that the larger the distance the higher the probability that the two random vectors are not identically distributed, i.e., $F_X \neq F_Y$.

Theorem 1. *Let* $X \sim F_X$ *and* $Y \sim F_Y$ *be two d-dimensional random vectors.* X', *and* Y' *are independent copies of* X *and* Y. *The corresponding characteristic functions of the two random vectors are* ϕ_X *and* ϕ_Y. *If* $0 < \alpha < 2$ *with* $\mathbb{E}||X||_2^\alpha < \infty$ *and* $\mathbb{E}||Y||_2^\alpha < \infty$ *then*

$$\int_{\mathbb{R}^d} \frac{|\phi_X(p) - \phi_Y(p)|^2}{||p||_2^{d+\alpha}} dp = W(d, \alpha) \mathcal{E}^\alpha(X, Y), \tag{1}$$

where

$$W(d, \alpha) = \frac{2\Pi^{\frac{d}{2}} \Gamma(1 - \frac{\alpha}{2})}{\alpha 2^\alpha \Gamma(\frac{\alpha+d}{2})}, \text{ with } \Gamma(\cdot) \text{ being the Gamma function,}$$

$$\mathcal{E}^\alpha(X, Y) = 2\mathbb{E}||X - Y||_2^\alpha - \mathbb{E}||X - X'||_2^\alpha - \mathbb{E}||Y - Y'||_2^\alpha. \tag{2}$$

Proof. See Lemma 1 in Appendix of Székely and Rizzo (2005).

Theorem 2. *Under assumptions of Theorem 1, $\mathcal{E}^\alpha(X, Y) = 0$ iff X and Y are identically distributed.*

Proof. See Theorem 2 (ii) in Székely and Rizzo (2005).

Therefore, the metric $\mathcal{E}^\alpha(X, Y)$ can be used to measure the divergence between two distributions. Let the samples of random vectors X, Y be $S_X = \{x_1, \ldots, x_m\}$ and $S_Y = \{y_1, \ldots, y_n\}$, respectively. The empirical counterpart of (2) replaces expectations by the averages and leads to

$$\hat{\mathcal{E}}^\alpha(S_X, S_Y) = \frac{mn}{m+n} \left(\frac{2}{mn} \sum_{i=1}^{m} \sum_{j=1}^{n} ||x_i - y_j||_2^\alpha \right.$$

$$\left. - \frac{1}{m^2} \sum_{i=1}^{m} \sum_{j=1}^{m} ||x_i - x_j||_2^\alpha - \frac{1}{n^2} \sum_{i=1}^{n} \sum_{j=1}^{n} ||y_i - y_j||_2^\alpha \right). \quad (3)$$

From Theorem 2 one sees that the larger the value of $\hat{\mathcal{E}}^\alpha(S_X, S_Y)$ the higher is the likelihood that the components in S_X, S_Y are from different distributions. Hence $\hat{\mathcal{E}}^\alpha(S_X, S_Y)$ can be used as the test statistics. Since the theoretical distribution of the test statistics $\hat{\mathcal{E}}^\alpha$ is intractable, a permutation test is employed under the assumption of independent random vectors. In order to accomplish this, $P = (m + n)!$ bootstrap samples $(B, C) = (b_1, \ldots, b_m, c_1, \ldots, c_n) = (a_{(1)}, \ldots, a_{(m+n)})$ can be generated by random shuffling of $\{a_1, \ldots, a_{m+n}; a_i = x_i, i \in 1, \ldots, m, a_i = y_{i-m}, i = m+1, \ldots, m+n\} = \{x_1, \ldots, x_m, y_1, \ldots, y_n; x_i, y_i \in \mathbb{R}^d\}$. For every permutation sample (B, C), the energy test statistics $\hat{\mathcal{E}}^\alpha(B, C)$ is calculated, which leads to a P-vector of test statistics based on P different permutation samples. It allows to compute the empirical distribution of $\hat{\mathcal{E}}^\alpha(S_X, S_Y)$. The critical value can be then obtained by choosing a quantile following the given confidence level. For more details on the permutation test and its related empirical distribution, please refer to Fisher (1937) and Pitman (1937).

2.2   Benchmark Tests

In this sub-section, tests used in three recent nonparametric control charts are briefly reviewed, including the SREWMA in Zou et al. (2012), the SMMST in Zhou et al. (2015) and the NPMVCP in M. Holland and Hawkins (2014). Control charts based on these tests are considered as benchmarks in later studies.

SREWMA. Proposed in Zou et al. (2012), a nonparametric multivariate EWMA control chart is constructed based on the spatial rank test to monitor the changes in the location parameter. It assumes that for a sequence of random vectors $X_{-g+1}, \ldots, X_0, X_1, \ldots, X_t \in \mathbb{R}^d$, where $X_{-g+1}, \ldots, X_0$ are g vectors before the starting point X_1, the multivariate change point problem is represented as:

$$X_i \overset{\text{i.i.d.}}{\sim} \begin{cases} \mu_0 + \Omega \varepsilon_i & \text{if } i \leq \tau, \\ \mu_1 + \Omega \varepsilon_i & \text{if } i > \tau, \end{cases} \quad (4)$$

where τ stands for the change index, and Ω is a full-rank $d \times d$ transformation matrix with the inverse $M = \Omega^{-1}$. It is assumed that $\varepsilon_i \in \mathbb{R}^d$ are i.i.d. with $\mathrm{Cov}(\varepsilon_i) = I_d$ and $\mathbb{E}(\varepsilon_i) = 0$. The test statistics and its asymptotic distribution are given as

$$Q_t^{R_E} = \frac{(2-\lambda)d}{\lambda \xi_t} V_t^\top V_t \to \chi_d^2, \ \lambda \to 0, \ \lambda t \to \infty, \tag{5}$$

with

$$V_t = (1-\lambda)V_{t-1} + \lambda R_E(\hat{M}_{t-1} X_t), \ V_0 = 0,$$
$$\xi_t = \hat{\mathbb{E}}\{R_F(MX_t)^\top R_F(MX_t)\},$$

where $\hat{M}_t = \hat{\Omega}_t^{-1}$, $R_E(X_t) = \frac{1}{g}\sum_{j=1}^g U(X_t - X_j)$ is the empirical version of the spatial rank for the d-vector X_t, and the theoretical counterpart is $R_F(X_t) = \mathbb{E}_{X_j}\{U(X_t - X_j)\}$ with U being the spatial sign function

$$U(X) = \begin{cases} (X^\top X)^{-1/2} X & \text{for } X = 0, \\ 0 & \text{else.} \end{cases} \tag{6}$$

SMMST. The multivariate version Wald-Wolfowitz runs test by Friedman and Rafsky (1979) has been integrated into the Hawkins et al. (2003) control chart based on the change point model in Hawkins et al. (2003) by Zhou et al. (2015) to perform nonparametric multivariate location surveillance. The main idea of the multivariate Wald-Wolfowitz runs test is to use the minimal spanning tree (MST) approach to generalize the sorted list in the univariate runs test to the multivariate context. That is, in the d-dimensional data set with $m + n$ observations $\{a_1, \ldots, a_{m+n}; a_i = x_i, i \in 1, \ldots, m, a_i = y_{i-m}, i = m+1, \ldots, m+n\} = \{x_1, \ldots, x_m, y_1, \ldots, y_n; x_i, y_i \in \mathbb{R}^d\}$ stemming from random vectors X and Y respectively, every observation is seen as a node and all the nodes can be connected by $(m+n)(m+n-1)/2$ edges. Friedman and Rafsky (1979) gives three steps to compute the test statistic.

1. Use the MST algorithm to construct the MST for all nodes in the data set, see Appendix in Friedman and Rafsky (1979).
2. Remove all edges, of which the two nodes are from different groups.
3. Compute the statistics $R = \#\{\text{disjoint sub-trees in the MST}\}$.

Furthermore, the change point problem in Zhou et al. (2015) is represented as

$$X_i \overset{\text{i.i.d.}}{\sim} \begin{cases} F_X & \text{if } i \le \tau, \\ F_Y & \text{if } i > \tau. \end{cases} \tag{7}$$

The null hypothesis $H_0 : F_X = F_Y$, will be rejected if R is smaller than a critical value, which follows the standard normal distribution as

$$W = \frac{R - \mathbb{E}(R)}{\sqrt{\mathrm{Var}(R|C)}} \to N(0,1), \text{if } m, n \to \infty,$$

where C is determined by the node degrees and the explicit formulas for expectation and variance can be found in Zhou et al. (2015).

NPMVCP. A nonparametric control chart using the multivariate rank based test by (Choi & Marden, 1997) is proposed in (M. Holland & Hawkins, 2014). It gives the multivariate change point model Hawkins et al. (2003) to identify changes in a sequence, $X_1, \ldots, X_t$, as follows

$$X_i \sim \begin{cases} F(\mu), & \text{if } i \leq \tau, \\ F(\mu + \delta), & \text{if } i > \tau, \end{cases} \tag{8}$$

and $H_0 : \delta = 0$, versus $H_1 : \delta \neq 0$. The test statistics and its asymptotic distribution are given for $k \in \{1, \ldots, t-1\}$ as:

$$\frac{tk}{t-k} \bar{r}_t^{(k)^\top} \tilde{\Sigma}_{k,t}^{-1} \bar{r}_t^{(k)} \to \chi_d^2, \text{ if } t \to \infty, \tag{9}$$

where $\tilde{\Sigma}_{k,t}$ is the pooled sample covariance matrix of the centered rank vector $\bar{r}_t^{(k)}$ computed using the kernel function. At last, M. Holland and Hawkins (2014) uses the test statistic

$$r_{k,t} = \bar{r}_t^{(k)^\top} \hat{\Sigma}_{k,t}^{-1} \bar{r}_t^{(k)},$$

where $\hat{\Sigma}_{k,t} = \left(\frac{t-k}{tk}\right)\hat{\Sigma}_t$ is the unpooled estimator of covariance matrix of centered ranks. Simulation study by M. Holland and Hawkins (2014) shows that the power of using pooled or unpooled estimator of covariance matrix leads to similar performance. However, for convenience of computation, the unpooled covariance estimator $\hat{\Sigma}_{k,t}$ is employed.

2.3   Phase I Change Point Model

Let $\{x_1, \ldots, x_T\}$ denote a sample of observations with the length of T. In Phase I detection without new observation, the detection is performed only based on the sample $\{x_1, \ldots, x_T\}$ as historical data. Hence, this type of change point analysis is retrospective and static. Phase I analysis has many applications in bio-statistics and transportation statistics (e.g., Székely & Rizzo, 2005; Matteson & James, 2014).

For the sake of simplicity, consider the case with only one change occurred at $\tau + 1$. Then the change point detection problem can be represented in the following test hypotheses,

$$H_0 : X_i \sim F_0, 1 \leq i \leq T,$$
$$H_1 : X_i \sim \begin{cases} F_0, 1 \leq i \leq \tau, \\ F_1, \tau + 1 \leq i \leq T. \end{cases}$$

A two-sample parametric or nonparametric test similar to those discussed in the previous section with the test statistics $B_{i,T}$ is usually applied here. Before conducting the permutation test, the significance level should be fixed. If $B_{i,T}$ is larger than a predefined critical value $h_{i,T}$, i.e., $B_{i,T} > h_{i,T}$, then the null hypothesis is rejected, meaning that the two sets of random vectors are not identically distributed. Then a detection point is admitted at i-th point. Since the change point location is unknown, hence the two-sample test will be performed at every point i, $1 \leq i < T$, i.e., conducting $T - 1$ dichotomizations. According to the change point model (Hawkins et al., 2003), the test statistics is derived from $B_{i,T}$, $i = 1, \ldots, T - 1$, as the largest value, such that

$$B_T = \max_{1 \leq i < T} B_{i,T}.$$

The null hypothesis is rejected if $B_T > h_T$, where h_T is the critical value derived from the distribution of B_T. The Type I error α in this context means that the model signals a change point when actually there is actually no change occurs. The distribution of the test statistic B_T can be obtained either by its asymptotic distribution (if available) or by simulation methods, e.g., permutation test. At the end, the location of the change can be estimated by

$$\hat{\tau} = \arg \max_{1 \leq i < T} B_{i,T}.$$

2.4   Phase II Change Point Model

In the contrary to the Phase I detection based on the fixed-sized sample $\{x_1, \ldots, x_T\}$, Phase II detection considers the dynamic sample $\{x_1, \ldots, x_t\}$ with an increasing size, i.e., the sample size t increases with time proceeding. For this reason, Phase II detection is also termed as online detection and sequential detection. For example, the stock price is updated with time, therefore the length of time series of returns is always increasing. Hence, the detection in Phase II concentrates on the dynamic stream data.

With the Phase I analysis in Section 2.3 at hand, Phase II can be extended from the Phase I to update the old sample size. That is, whenever a new observation x_t arrives, a new sample $\{x_1, \ldots, x_T, x_{T+1}, \ldots, x_t\}$ is constructed and the new sample size is denoted here as t. For example, if the old sample is $\{x_1, \ldots, x_T\}$ with $t = T$ and the new arrival is x_{T+1}, then the new sample becomes $\{x_1, \ldots, x_T, x_{T+1}\}$ and t becomes $t = T + 1$. For every new arrival of observation, the Phase I analysis will be performed based on the new sample $\{x_1, \ldots, x_T, x_{T+1}, \ldots, x_t\}$. For this sample, $t - 1$ two-sample tests will be performed and computed. Further, $B_t = \max\{B_{1,t}, \ldots, B_{t-1,t}\}$. Hence the null hypothesis is rejected if $B_t > h_t$. The Type I error α can be thus represented with

$$\mathbb{P}(B_1 > h_1) = \alpha, \ t = 1,$$
$$\mathbb{P}(B_t > h_t | B_{t-1} \leq h_{t-1}, \ldots, B_1 \leq h_1) = \alpha, \ t > 1. \tag{10}$$

In statistical process control, the in-control average run length (ARL_0), is the inverse of the Type I error, i.e., $ARL_0 = 1/\alpha$, which stands for the average step length of the detection until the first erroneous alarm signals.

3 Simulation Study

In the study of statistical process monitoring, the assessment of change-point detection methods uses mainly two measures, the in-control average run length (ARL_0) and the out-of-control average run length (ARL_1). ARL_0 assumes that the time series follows a distribution without changes in order to calculate the steps until the first erroneous signal flags. Therefore the larger the ARL_0 the better the model. ARL_1 assumes that the process has a change point in a known point in order to compute the average run length until the model detects this pre-set change. Since there is a delay in detection, the detection method, therefore, is expected to have a small ARL_1 value.

The recent paper studying nonparametric multivariate control chart using the change point model (Hawkins et al., 2003) is NPMVCP in M. Holland and Hawkins (2014). Therefore, we choose NPMVCP as the benchmark model for comparison in this paper, which is a mainstream nonparametric control chart for multivariate location shift detection. Since this paper used the code provided in R package NPMVCP in M. Holland and Hawkins (2014) without the usage of optimal quarantine technique, for fair comparison, the quarantine was not considered for both models. Here, the in-control length (ICL) is set as 32, and out-of-control length (OCL) is set separately as 100 and 200, consistent to the default set-up in NPMVCP. Choosing the warm-up equal to 32, because firstly the defualt set-up in NPMVCP is fixed in 32 and secondly 32 is short for re-starting, the control chart is especially crucial in finanical surveillance.

As the test integrated in the ETCC is based on permutation samples, hence the choice of the number of simulation runs should be considered. As all the metrics are computed based on the i.i.d. samples, the mean of ARL_1s will converge under the law of large numbers. In order to choose an appropriate size of simulation, a simple simulation study was conducted. The DGP is a five dimensional standard Gaussian distribution, $N_5(0, \mathcal{I})$, shifted to $N_5(3, \mathcal{I})$, where $\mathcal{I}$ is the identity matrix and the warm-up is set to $\tau = 32$ identical to the setting in the package NPMVCP. As can be seen in Figure 1, the simulation runs larger than 50 led to the similar results and the mean of both control charts' ARL_1s arrived close to the run of 50. Hence in this paper, the simulation size was chosen as 50 runs for sufficiency.

In the next three scenarios, we consider shifts in mean (whole vector and single constituent) and variance, and compare the performance with NPMVCP. In the fourth scenario with mean shift, we compare ETCC with SMMST and SREWMA.

a) In the mean shift scenario, the detection assessment sets the break of $\tau = 32$, i.e., ICL = 32 and OCL $\in \{100, 200\}$, and the distributions used in simulation are $N(0, \mathcal{I})$, Student-$t_5(0, \mathcal{I})$ and Laplace$(0, \Sigma_L)$, $\Sigma_L = (a_{ij})$, $a_{ii} = 11$, $a_{ij} = 10$. The shifts are set as $\delta = 0, 0.25, 0.5, 0.75, 1, \ldots, 9$. Hence, the in-control distributions, $N(0, \mathcal{I})$, Student-$t_5(0, \mathcal{I})$ and Laplace$(0, \Sigma_L)$, will shift to $N(\delta, \mathcal{I})$, Student-$t_5(\delta, \mathcal{I})$ and Laplace(δ, Σ_L) at the 32nd observation. To emphasize the performance of the ETCC under the in-control situation, we give, in Table 1, ARL_0s for ETCC and NPMVCP with corresponding empirical standard deviations computed over the simulation runs. As can be seen ARL_0s for ETCC are almost equal in OCL (100 and 200) where ARL_0s for NPMVCP are twice smaller with often more than twice bigger variance. The ARL performance (ARL_0 for $\delta = 0$ and ARL_1 else) is shown in Figure 2. For all three distributions, the ETCC performs better in moderate to large shifts ($\delta \geq 2$) for three dimensional cases and in small to large shifts ($\delta \geq 0.75$) in ten dimensional cases, see Gaussian and t_5. With the increase of the dimension of data, the performance of ETCC is steadily improving.

b) In the scenario of the single component mean shift, the breaks are set at $\tau = 32$, i.e., ICL $= 32$ and OCL $\in \{100, 200\}$, and the distributions used in simulation are $N(0, \mathcal{I})$, Student-$t_5(0, \mathcal{I})$ and Laplace$(0, \Sigma_L)$, $\Sigma_L = (a_{ij})$, $a_{ii} = 11$, $a_{ij} = 10$. The shifts are set as $\delta = 0, 0.25, 0.5, 0.75, 1, \ldots, 9$. The shifting method here is similar to the one in the first scenario, but only the last column shifts by δ while the other columns are kept unchanged $N((0, \ldots, 0, \delta)^\top, \mathcal{I})$, Student-$t_5((0, \ldots, 0, \delta)^\top, \mathcal{I})$ and Laplace$((0, \ldots, 0, \delta)^\top, \Sigma_L)$. The single component mean shift scenario shows that NPMVCP performs well in small shifts and the ETCC performs well in moderate shift ($\delta \geq 2$), see Figure 3. However, in all the categories, the ETCC outperforms the NPMVCP in ARL_0. The NPMVCP has only roughly 60 percent correct detection, which is far worse than the ETCC, where the disadvantageous performance of NPMVCP is consistent with the result in M. Holland and Hawkins (2014). According to Table 1 and the above analysis, one can conclude that the ETCC in mean shift detection is capable and robust.

c) In the covariance shift part, the DGPs are set as the $N(0, \mathcal{I})$ and Student-t_5 with $\sigma^2 = 0.25, 0.5, 0.75, 1, 2, \ldots, 11$. Here it means there is no change when $\sigma^2 = 1$. Hence the in-control distributions will shift to $N(0, \mathcal{I}\sigma^2)$ and Student-$t_5(0, \mathcal{I}\sigma^2)$. The ETCC outperforms the NPMVCP in most cases, see Figure 4, while NPMVCP has the ability to detect the small covariance shift, e.g. in scale of $\sigma^2 = 2$. In larger covariance shifts or larger dimension data sets, the ETCC gave better results. The NPMVCP shows to be robust to the changes of dimensions or distributions, while the ETCC shows high sensitivity to the increase of dimension, and ARL_1 strongly decrease with the dimension.

d) Additionally, as mentioned in Section 2.2, the ETCC is compared with another two nonparametric control charts, namely the SMMST and the SREWMA in scenario of 200 ARL_1-steps mean shift under Gaussian, t_5 and Gamma$_5$. Breaks τ are set as 40 and 90, and shifts $\delta = 1, 1.5, 2, 3, 4$. The results of SMMST and SREWMA are collected from the Tables 2, 3, 4, 5 in Zhou et al. (2015). Using the same simulation setting in Zhou et al. (2015), we tested the performance of ETCC. Since our simulation based on independent samples hence for convenience, we did not re-run the SMMST and SREWMA but just took the result from Zhou et al. (2015). In order to further support the robustness and capacity of the ETCC, Figure 5 provides another evidence. The ETCC performs generally better than the other benchmarks, especially in Gaussian and t_5 cases.

4   Conclusion

This paper proposes a nonparametric multivariate control chart to detect the multiple change points in high-dimensional stream data. It has four features. Firstly, it is a nonparametric control chart requiring no assumption on the process, compared with the classical parametric control chart. Secondly, it is oriented to Phase II change point detection that is central for real time surveillance of stream data and can be applied extensively, e.g., in industrial quality control, finance, medical science, and geology. Thirdly, the control charts is designed for multivariate time series, which is more practical and informative for catching the essence of data as a whole than uni-variate time series. Last but the most important feature of the ETCCs is that it monitors not only mean or covariance, but monitors mean and covariance simultaneously. In that simulation study, the mean and covariance shifts were investigated and the control chart has shown outstanding performance compared to the benchmark models.

References

Chakraborti, S., Qiu, P., & Mukherjee, A. (2015). Editorial to the special issue: Nonparametric statistical process control charts. *Quality and Reliability Engineering International*, *31*(1), 1–2.

Choi, K., & Marden, J. (1997). An approach to multivariate rank tests in multivariate analysis of variance. *Journal of the American Statistical Association*, *92*(440), 1581–1590.

Crosier, R. B. (1988). Multivariate generalizations of cumulative sum quality-control schemes. *Technometrics*, *30*(3), 291-303.

Erdman, C., Emerson, J. W., et al. (2007). bcp: an R package for performing a Bayesian analysis of change point problems. *Journal of Statistical Software*, *23*(3), 1–13.

Fisher, R. A. (1937). *The design of experiments*. Oliver And Boyd; Edinburgh; London.

Friedman, J. H., & Rafsky, L. C. (1979). Multivariate generalizations of the wald-wolfowitz and smirnov two-sample tests. *The Annals of Statistics*, 697–717.

Hawkins, D. M., & Deng, Q. (2010). A nonparametric change-point control chart. *Journal of Quality Technology*, 165-173.

Hawkins, D. M., Qiu, P., & Kang, C. W. (2003). The changepoint model for statistical process control. *Journal of Quality Technology*, *35*(4), 355.

Hawkins, D. M., & Zamba, K. (2005a). A change-point model for a shift in variance. *Journal of Quality Technology*, *37*(1), 21.

Hawkins, D. M., & Zamba, K. (2005b). Statistical process control for shifts in mean or variance using a changepoint formulation. *Technometrics*, *47*(2), 164–173.

Holland, M., & Hawkins, D. (2014). A control chart based on a nonparametric multivariate change-point model. *Journal of Quality Technology*, *46*, 1975-1987.

Holland, M. D. (2013). NPMVCP: Nonparametric multivariate change point model. *Reference manual.* Retrieved from ftp://cran.r-project.org/pub/R/web/packages/NPMVCP/index.html

Hosking, J. R. (1980). The multivariate portmanteau statistic. *Journal of the American Statistical Association, 75*(371), 602–608.

James, N., & Matteson, D. (2015). ecp: An R package for nonparametric multiple change point analysis of multivariate data. *Journal of Statistical Software, 62*(1), 1-25.

Killick, R., & Eckley, I. (2011). Changepoint analysis with the changepoint package in R. In *The R user conference, useR! 2011 august 16-18 2011 university of warwick, coventry, uk* (p. 51).

Kim, A. Y., Marzban, C., Percival, D. B., & Stuetzle, W. (2009). Using labeled data to evaluate change detectors in a multivariate streaming environment. *Signal Processing, 89,* 2529-2536.

Knoth, S. (2016). spc: Statistical process control - collection of some useful functions. *Reference manual.* Retrieved from https://cran.r-project.org/web/packages/spc/index.html

Lowry, C. A., Woodall, W. H., Champ, C. W., & Rigdon, S. E. (1992). A multivariate exponentially weighted moving average control chart. *Technometrics, 34*(1), 46-53.

Matteson, D. S., & James, N. A. (2014). A nonparametric approach for multiple change point analysis of multivariate data. *Journal of the American Statistical Association, 109*(505), 334-345.

Page, E. S. (1954). Continuous inspection schemes. *Biometrika, 41*(1/2), 100–115.

Pitman, E. J. (1937). Significance tests which may be applied to samples from any populations. *Supplement to the Journal of the Royal Statistical Society, 4*(1), 119–130.

Qiu, P. (2018). Some perspectives on nonparametric statistical process control. *Journal of Quality Technology, 50*(1), 49–65.

Qiu, P., & Hawkins, D. (2001). A rank-based multivariate cusum procedure. *Technometrics, 43*(2), 120-132.

Qiu, P., & Hawkins, D. (2003). A nonparametric multivariate cumulative sum procedure for detecting shifts in all directions. *Journal of the Royal Statistical Society: Series D (The Statistician), 52*(2), 151-164.

Rizzo, M. L., & Székely, G. J. (2016). Package 'energy'. *R User's Manual.*

Roberts, S. (1959). Control chart tests based on geometric moving averages. *Technometrics, 1*(3), 239–250.

Ross, G. J., et al. (2013). Parametric and nonparametric sequential change detection in R: The cpm package. *Journal of Statistical Software, 78.*

Serfling, R. J. (2009). *Approximation theorems of mathematical statistics* (Vol. 162). John Wiley & Sons.

Shewhart, W. A. (1931). *Economic control of quality of manufactured product.* ASQ Quality Press.

Shewhart, W. A., & Deming, W. E. (1939). *Statistical method from the viewpoint of quality control*. Courier Corporation.

Sims, C. A. (1980). Macroeconomics and reality. *Econometrica: Journal of the Econometric Society*, 1–48.

Székely, G. J., & Rizzo, M. L. (2004). Testing for equal distributions in high dimension. *InterStat*, *5*.

Székely, G. J., & Rizzo, M. L. (2005). Hierarchical clustering via joint between-within distances: Extending Ward's minimum variance method. *Journal of Classification*, *22*(2), 151-183.

Székely, G. J., & Rizzo, M. L. (2013). Energy statistics: statistics based on distances. *Signal Processing*, *143*, 1249-1272.

Woodall, W. H., & Montgomery, D. C. (2014). Some current directions in the theory and application of statistical process monitoring. *Journal of Quality Technology*, *46*(1), 78.

Xu, Y. F. (2017). Energyonlinecpm: Distribution free multivariate control chart based on energy test [Computer software manual]. Retrieved from https://cran.r-project.org/web/packages/EnergyOnlineCPM/index.html

Zech, G., & Aslan, B. (2003). A multivariate two-sample test based on the concept of minimum energy. *Proc. Statistical Problems in Particle Physics, Astrophysics, and Cosmology*, 8–11.

Zeileis, A., Leisch, F., Kleiber, C., & Hornik, K. (2005). Monitoring structural change in dynamic econometric models. *Journal of Applied Econometrics*, *20*(1), 99–121.

Zhou, M., Zi, X., Geng, W., & Li, Z. (2015). A distribution-free multivariate change-point model for statistical process control. *Communications in Statistics: Simulation and Computation*, *44*, 1975-1987.

Zou, C., & Tsung, F. (2011). A multivariate sign EWMA control chart. *Technometrics*, *53*, 84-97.

Zou, C., Wang, Z., & Tsung, F. (2012). A spatial rank-based multivariate ewma control chart. *Naval Research Logistics (NRL)*, *59*(2), 91–110.

Appendix: Tables And Figures of Simulation Results

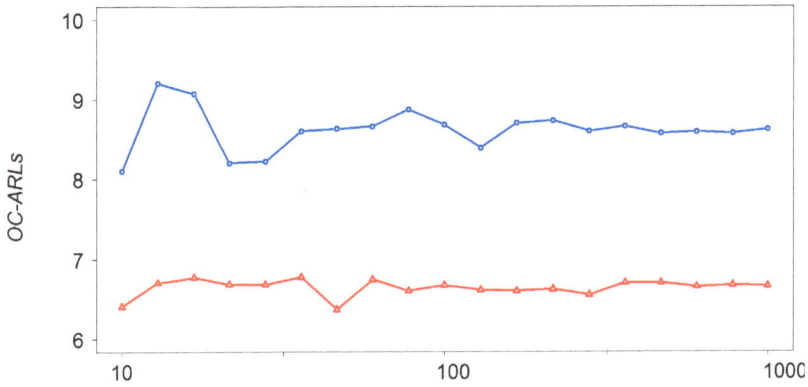

Fig. 1. Comparison of ARL_1s under NPMVCP and ETCC through 10 to 1000 runs of simulation.

Table 1. Comparison of ETCC against the NPMVCP model M. Holland and Hawkins (2014) in In-Control ARL for mean shift with 100 and 200 ARL_1-steps (with standard deviations in parentheses).

		Gaussian		t		Laplace	
ARL_0	Dim.	ETCC	NPMVCP	ETCC	NPMVCP	ETCC	NPMVCP
200	3	182.36 (50.29)	124.82 (71.55)	182.56 (48.89)	118.66 (74.11)	187.84 (41.96)	135.62 (67.62)
	10	195.46 (19.71)	138.62 (70.29)	179.26 (52.06)	135.02 (69.45)	183.47 (45.77)	140.28 (67.84)
100	3	95.54 (16.91)	67.30 (40.25)	90.07 (27.13)	68.00 (34.34)	93.30 (20.45)	62.84 (35.35)
	10	91.13 (24.29)	58.24 (38.50)	88.38 (29.25)	74.12 (34.49)	93.98 (21.11)	69.34 (34.68)

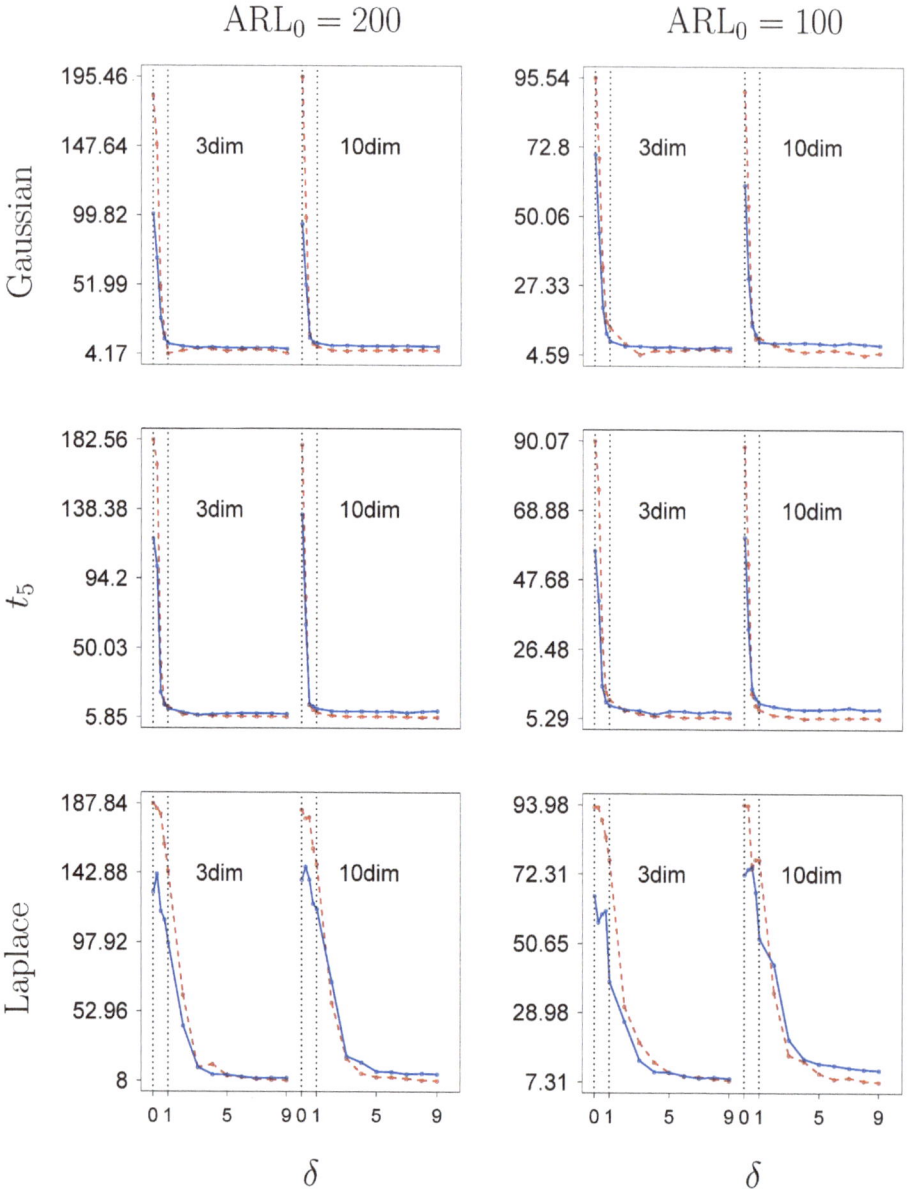

Fig. 2. Simulation results for mean shift with DGPs of Gaussian, Student-t_5 and Laplace distributions. The blue line stands for NPMVCP and the red for ETCC. The data in the figure are given in Table 2.

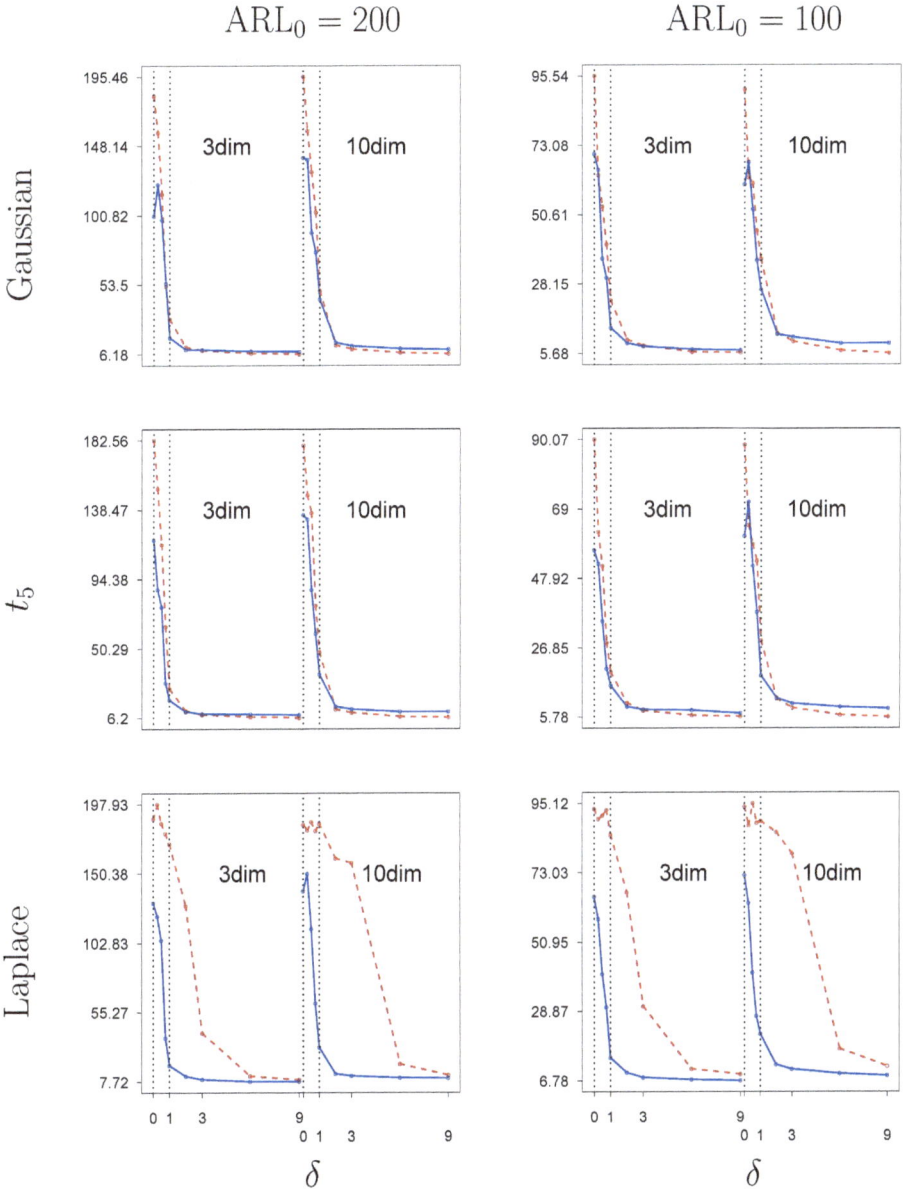

Fig. 3. Single mean shift for multivariate Gaussian, Student-t_5 and Laplace with mean $\mu_k + \delta$, $\delta \in \{0, 0.25, 0.50, 0.75, 1, 2, 3, 6, 9\}$. The red line stands for the ETCC and the blue line for the M. Holland and Hawkins (2014).

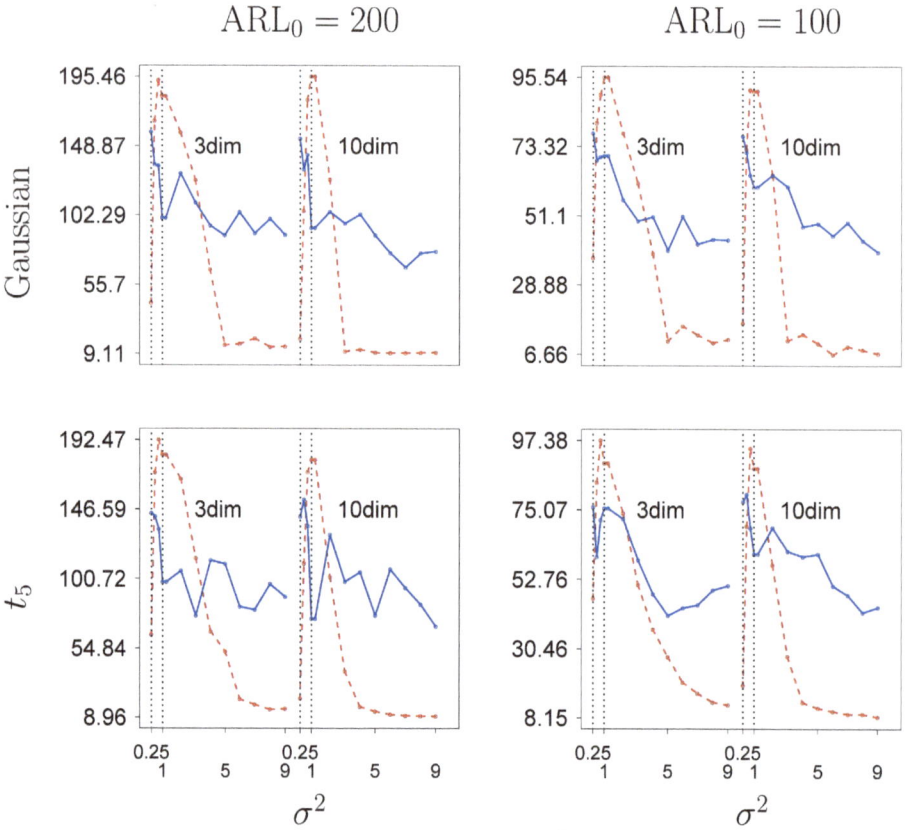

Fig. 4. Simulation results for covariance shift with DGPs of Gaussian and Student-t_5. The blue line stands for NPMVCP and the red line for ETCC. The data used in the figure are given in Table 3.

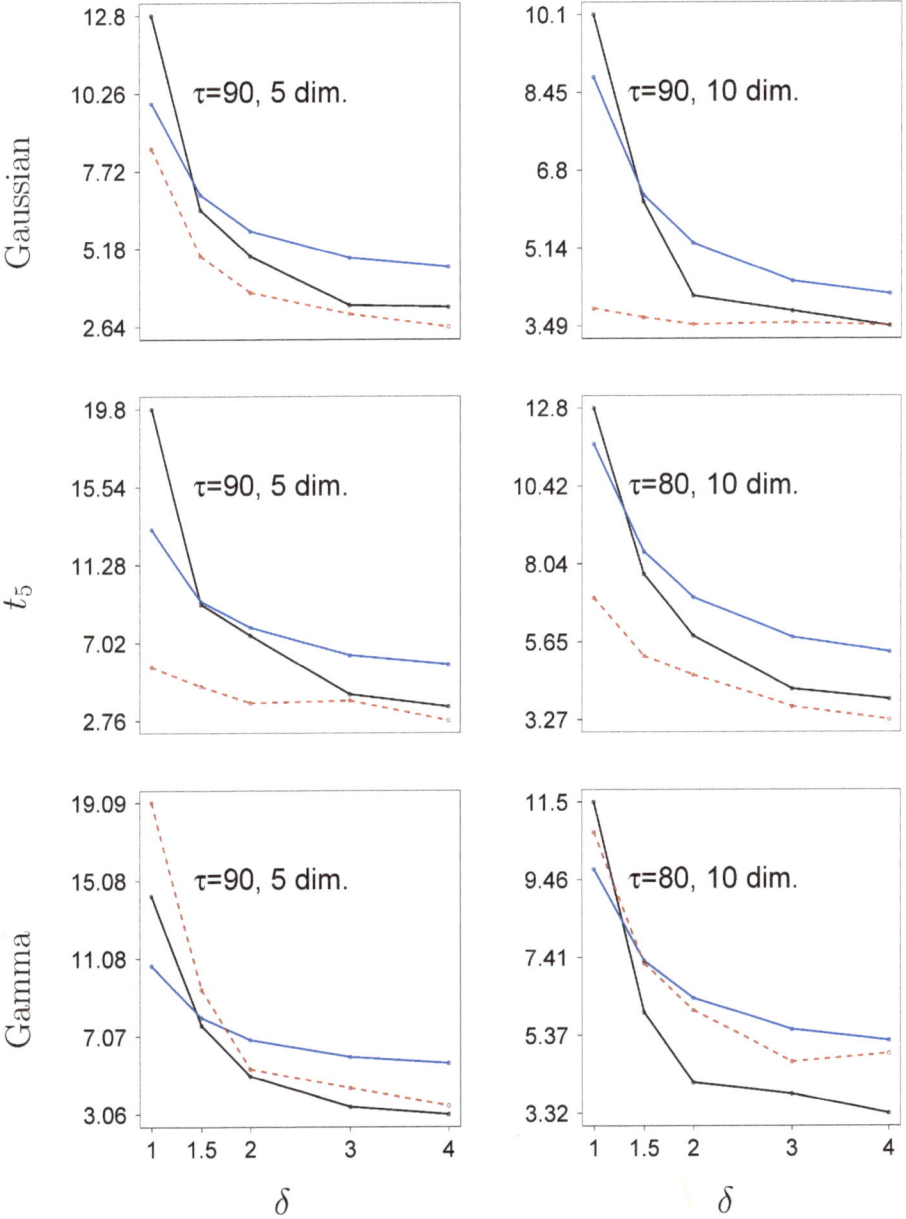

Fig. 5. Comparison of simulation results of the ETCC (red) with SMMST (Zhou et al., 2015) in black and SREWMA Zou et al. (2012) in blue appeared in Tables 2, 3, 4 in Zhou et al. (2015). The data used in the figure are given in Table 4.

Table 2. Mean shift in standard Gaussian, Student-t_5 and Laplace cases. The outperformed points of NPMVCP compared with the ETCC are in bold. The in-control length is set as 32 and out-of-control as 100 and 200, and change point $\tau = 32$.

Dimensions	δ	Mean Gaussian Shift $ARL_o=200$ ETCC	NPMVCP	$ARL_o=100$ ETCC	NPMVCP	Mean t Shift $ARL_o=200$ ETCC	NPMVCP	$ARL_o=100$ ETCC	NPMVCP	Mean Laplace Shift $ARL_o=200$ ETCC	NPMVCP	$ARL_o=100$ ETCC	NPMVCP
3	0	182.36	100.48	95.54	70.24	182.56	98.46	90.07	75.50	187.84	114.12	93.30	65.36
	0.25	148.75	**70.12**	69.06	**44.54**	166.55	**102.04**	75.20	**41.26**	184.58	**141.88**	93.22	**57.22**
	0.50	50.69	**28.62**	33.40	**20.08**	40.31	**21.86**	29.82	**15.28**	181.26	**117.92**	89.22	**59.66**
	0.75	17.82	**14.42**	15.58	**11.64**	14.45	**14.02**	13.29	**10.36**	161.56	**112.22**	83.76	**60.74**
	1	4.17	10.86	13.53	**9.06**	12.63	**11.10**	10.90	**9.18**	143.72	**97.40**	76.62	**38.50**
	2	6.18	9.20	8.31	**7.46**	8.84	8.84	7.65	8.00	63.46	**43.54**	30.78	**26.10**
	3	7.97	8.20	4.67	7.48	7.00	7.20	6.66	7.64	16.62	16.66	19.58	**13.98**
	4	7.27	8.64	6.00	7.04	7.48	7.56	5.90	6.50	18.61	**11.98**	13.42	**10.40**
	5	5.95	8.06	5.65	7.18	6.63	7.98	6.03	7.42	11.23	11.78	10.17	10.26
	6	6.89	7.94	6.22	6.92	6.33	8.22	5.43	7.40	10.28	10.54	9.05	9.16
	7	7.08	8.28	6.51	6.72	6.15	8.22	5.60	6.90	9.12	9.60	8.81	**8.44**
	8	6.65	8.24	6.29	7.16	6.18	8.18	5.59	7.48	8.60	9.82	8.27	8.78
	9	4.48	7.56	5.95	6.96	5.98	7.78	5.46	7.08	8.27	9.78	7.71	8.34
10	0	195.46	93.74	91.13	60.28	179.26	74.28	88.38	60.72	183.77	99.64	93.98	65.86
	0.25	98.29	**52.02**	53.54	**29.76**	82.53	**64.86**	52.45	**32.82**	178.21	146.68	93.73	**73.82**
	0.50	17.74	**15.22**	15.51	**14.34**	13.68	14.22	12.86	**14.44**	178.88	138.26	73.94	**74.44**
	0.75	10.90	12.18	9.94	**11.42**	10.52	12.64	9.25	14.44	158.58	122.72	76.83	**66.64**
	1	8.97	11.46	10.27	**8.88**	8.68	11.26	7.97	9.94	148.20	119.28	76.71	**52.14**
	2	6.56	9.70	7.90	8.54	6.81	9.76	6.26	8.94	58.41	71.94	35.18	43.94
	3	6.09	9.94	6.38	8.56	6.27	9.40	6.01	8.24	22.38	23.90	15.70	20.54
	4	6.65	9.52	5.59	8.70	6.21	9.68	5.29	8.02	12.68	19.88	13.76	14.56
	5	6.35	9.58	6.01	8.42	6.17	9.36	5.44	8.06	10.41	13.98	9.97	13.04
	6	6.76	9.64	6.23	8.08	6.20	9.50	5.34	8.16	10.23	13.70	8.27	12.52
	7	6.75	9.72	6.05	8.08	6.05	8.78	5.43	8.60	9.37	12.38	8.64	11.74
	8	6.73	9.42	4.59	8.20	5.87	9.56	5.61	8.00	8.38	12.76	7.66	11.26
	9	6.35	9.30	5.34	7.84	5.85	9.92	5.31	8.10	8.00	12.30	7.31	10.98

Table 3. Variance shift in Gaussian and Student-t_5 cases. The outperformed points of NPMVCP compared with the ETCC are in bold. The in-control length is set as 32 and out-of-control as 100 and 200, and change point $\tau = 32$.

		Gaussian Variance Shift				t Variance Shift			
		ARL_o=200		ARL_o=100		ARL_o=200		ARL_o=100	
Dimensions	σ^2	ETCC	NPMVCP	ETCC	NPMVCP	ETCC	NPMVCP	ETCC	NPMVCP
3	0.25	43.24	158.30	37.66	77.50	63.62	143.94	46.64	75.90
	0.50	165.77	**136.22**	80.88	**68.80**	170.91	**141.86**	83.84	**59.78**
	0.75	192.96	**135.38**	89.96	**69.94**	192.47	**133.46**	97.38	**71.70**
	2	157.84	**130.32**	77.56	**56.12**	166.64	**105.92**	73.98	**72.10**
	3	125.76	**110.62**	61.44	**49.40**	114.27	**76.34**	50.98	58.78
	4	65.78	95.38	39.02	50.68	65.87	112.88	36.70	48.00
	5	14.92	88.60	10.91	40.00	52.58	110.50	27.85	41.10
	6	15.63	104.22	15.72	50.86	21.24	82.26	19.66	43.52
	7	19.32	90.08	12.94	42.06	17.38	80.22	16.02	44.50
	8	13.65	99.78	10.35	43.52	14.16	97.18	13.25	49.30
	9	13.88	89.04	11.40	43.32	14.54	88.78	12.31	50.72
	10	13.84	89.74	12.01	44.82	12.31	59.92	12.19	49.10
	11	8.95	82.02	16.02	41.70	12.14	89.34	10.96	38.24
10	0.25	19.06	153.84	16.58	76.56	21.48	141.76	18.64	77.62
	0.50	105.14	133.10	72.38	71.44	110.68	153.04	69.84	80.04
	0.75	179.82	**142.72**	91.56	**64.14**	171.08	**135.62**	94.78	**69.20**
	2	126.20	**104.52**	64.02	64.16	101.82	129.78	57.32	69.36
	3	10.66	96.72	11.10	60.38	39.60	98.82	28.06	61.66
	4	11.96	102.84	13.04	47.60	15.98	105.12	13.14	59.94
	5	9.91	88.72	10.15	48.60	12.86	76.44	11.32	60.72
	6	9.61	76.84	6.66	44.72	10.88	106.94	10.25	50.64
	7	9.61	67.26	9.19	48.92	10.15	94.66	9.37	47.60
	8	9.91	76.82	8.09	43.16	9.83	83.68	9.40	42.10
	9	10.02	77.98	6.90	39.48	9.73	69.28	8.46	43.76
	10	9.11	74.90	10.36	39.38	8.96	90.80	8.15	47.56
	11	6.49	85.68	6.71	40.78	8.94	77.72	7.78	45.74

Table 4. Out-of-control ARLs' comparison between the ETCC and the SMMST and the SREWMA control charts in context of Gaussian, t_5 and $Gamma_3$ mean shift. The performance of the SMMST and the SREWMA control charts is based on the Table 2, 3, 4, 5 in Zhou et al. (2015).

		Gaussian						t_5			$Gamma_5$			Mix		
		SMMST		SREWMA		ETCC		SMMST	SREWMA	ETCC	SMMST	SREWMA	ETCC	SMMST	SREWMA	ETCC
dim	δ	$\tau=40$	$\tau=90$	$\tau=40$	$\tau=90$	$\tau=40$	$\tau=90$	$\tau=90$	$\tau=90$	$\tau=90$	$\tau=90$	$\tau=90$	$\tau=90$	$\tau=90$	$\tau=90$	$\tau=90$
5	1	14.20	12.80	11.40	9.93	8.76	8.45	19.80	13.20	5.69	14.30	10.70	19.09	19.80	12.40	16.27
	1.5	7.35	6.46	7.69	6.95	5.26	4.94	9.13	9.28	4.63	7.64	8.03	9.48	8.91	9.04	9.72
	2	4.97	4.95	6.39	5.77	4.06	3.76	7.43	7.87	3.73	5.01	6.89	5.38	7.06	7.37	6.75
	3	3.59	3.35	5.42	4.89	3.20	3.06	4.20	6.35	3.85	3.45	6.01	4.40	4.50	6.18	5.03
	4	3.38	3.29	5.10	4.59	3.22	2.64	3.52	5.84	2.76	3.06	5.69	3.50	3.48	5.74	3.44
		$\tau=40$	$\tau=90$	$\tau=40$	$\tau=90$	$\tau=40$	$\tau=90$	$\tau=80$	$\tau=80$	$\tau=80$	$\tau=80$	$\tau=80$	$\tau=80$	$\tau=80$	$\tau=80$	$\tau=80$
10	1	11.00	10.10	9.57	8.77	4.96	3.66	12.80	11.70	4.22	11.50	9.73	10.70	18.20	10.00	13.25
	1.5	6.26	6.13	6.80	6.27	4.98	3.80	7.72	8.41	5.58	5.97	7.32	7.25	7.36	7.48	8.15
	2	4.27	4.13	5.73	5.25	4.78	3.88	5.84	7.01	4.38	4.13	6.34	6.02	5.78	6.34	6.71
	3	3.85	3.81	4.89	4.44	4.82	3.76	4.21	5.80	3.60	3.83	5.52	4.67	3.90	5.47	4.96
	4	3.61	3.49	4.60	4.17	4.72	3.32	3.90	5.35	3.40	3.32	5.23	4.89	3.74	5.13	4.17

WeibullR: An R Package for Weibull Analysis for Reliability Engineers

David J. Silkworth

OpenReliability.org
djsilk@openreliability.org

Abstract. The `WeibullR` package provides a flexible data entry capability with three levels of usage. Quick Fit Functions, `wblr` object model, and technical back end functions. `WeibullR` should appeal to the newest practitioners to the `R` community as well as seasoned researchers willing to examine deeper aspects of analysis.

Keywords: Survival · Reliability · Confidence.

DOI: 10.35566/isdsa2019c3

1 Summary

`WeibullR` intends to provide a complete user friendly application in the `R` environment. While there is little novel technology presented for life data analysis, the guiding design features include a flexible data entry capability, calculation performance and an attractive graphical presentation.

The target audience is intended to be engineering and manufacturing practitioners, who might be introduced to `R` for the first time. For this reason, package dependencies have been avoided so that installation is simple and the instruction set does not appear too demanding for newcomers.

Performance is enhanced by use of C++ code by way of the `Rcpp` and `RcppArmadillo` libraries, which are silently imported on installation and load. The C++ code greatly improves the speed of several complex looped calculations such as pivotal analysis, likelihood profiles, and third parameter optimizations. Stability of these calculations is monitored and can be improved due to the hand crafting of key optimization routines.

1.1 Quick Fit functions

Much of the detailed code required to generate regression lines, and confidence bounds on a plot with an informative legend are encapsulated in easy to use functions. The first set of functions to introduce are Quick Fit functions. Placing characteristics of the fit in the function name and use of reasonably expected defaults enables easy access to a complete analysis in one simple command line. The user is immediately rewarded with the graphic presentation in Figure 1.

```
MRRln2p(gears,T,T)
```

Files for reproducing all figures including data and code are available at: https://github.com/openrelia/ISDSA/blob/master/Figures/.

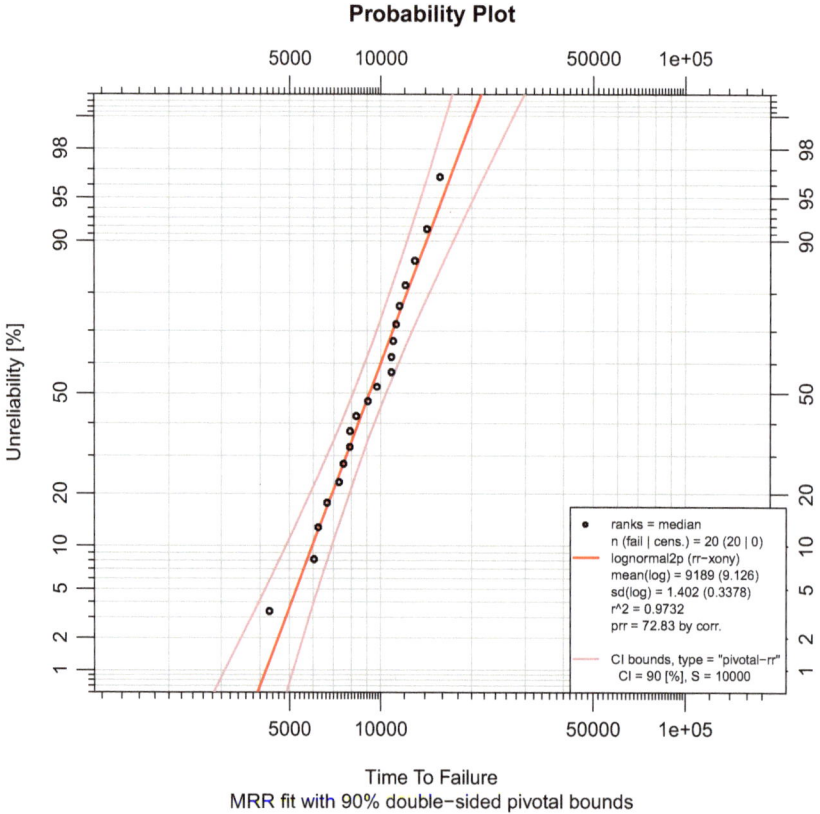

Fig. 1. A Quick Fit Probability Plot

1.2 The `wblr` Object Model

The `wblr` object model provides the user much more control over the analysis and plot generation. An object is created by passing the data into the `wblr` command.

```
my_object <- wblr(F3.13da)
```

Then the object is then modified by a fit specification.

```
my_object <- wblr.fit(my_object, dist="lognormal", col="magenta")
```

A single object may contain multiple fits, so we can provide additional fit specifications.

```
my_object <- wblr.fit(my_object, dist="weibull2p", col="blue")
```

```
my_object <- wblr.fit(my_object, dist="weibull3p", col="red")
```

The `wblr` object is a registered S3 object for plotting, so the simple command:

```
plot(my_object, main=("Multi-distribution Plot"))
```

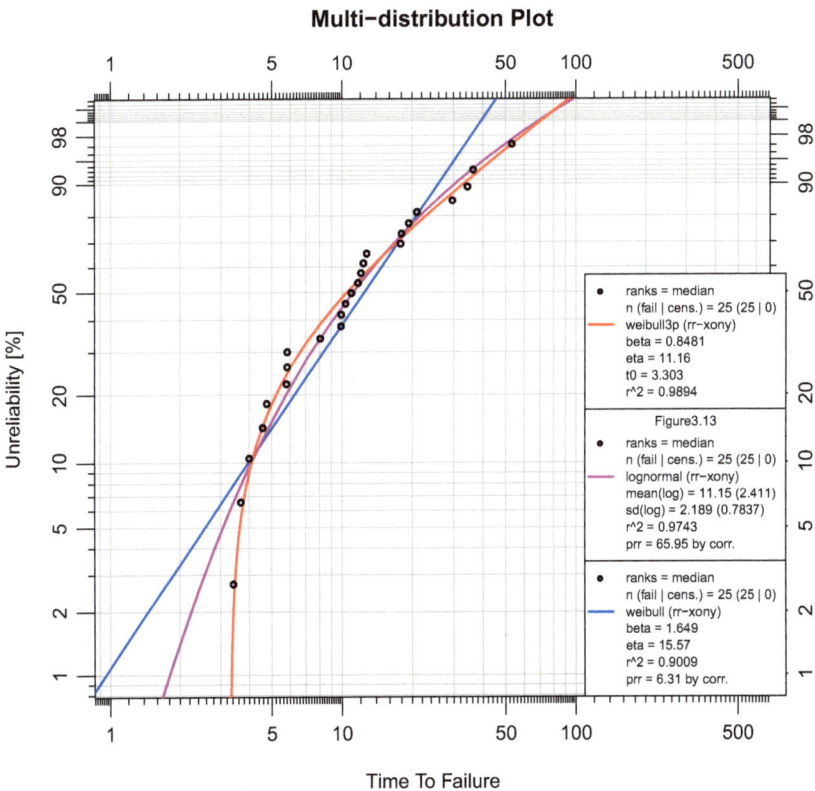

Fig. 2. A `wblr` Object Plot

2   Confidence Interval Bounds

`WeibullR` implements four types of confidence interval bound calculations:

- Beta binomial bounds are non-parametric and can be placed on any fit.

- Pivitol bounds are formed by way of a parametric bootstrap on rank regression fitting.
- For maximum likelihood estimation the classical normal-approximation bounds are formed utilizing solution of the negative hessian matrix (second derivatives of Fisher information matrix).
- A more complex calculation is also implemented based on the likelihood profile, also known as likelihood ratio bounds.

2.1   Likelihood Contours

Of the four confidence bound methods implemented in `WeibullR` likelihood ratio bounds are considered to be most authoritative, yet are more challenging to calculate. The precursor calculation for likelihood ratio bounds is the generation of a likelihood contour at specified confidence level for a given model. Contours are horizontal slices through the likelihood mound that is peaked by the maximum likelihood estimate. Contour slices are made at ratio values based on the relationship:

```
ratioLL  =  MLLx - qchisq(CL, dof)/2
```

Where the ratio is formed by subtraction of log-likelihood values. `MLLx` is the maximum log-likelihood estimate, `CL` is the confidence limit, and dof the degrees of freedom. Degrees of freedom are 1 for comparison of the model fit itself, or 2 when comparison is made against other data.

Using S3 registration of R's core contour function a contour map can easily be generated from a `wblr` object. Here a famous dataset for comparisons found as `MASS::gehan` is used to compare an early chemotherapy drug 6-mp against a control.

```
obj1 <- wblr(control, col="orange")
obj2 <- wblr(treat6mp, col="purple")
contour.wblr(list(obj1, obj2))
```

The code above has been abbreviated to demonstrate the simplicity of `WeibullR`. Expanded scripts for the data handling and plot annotations is provided in the `WeibullR.gallery` on Github as presented in Appendix B.

A particularly challenging contour calculation was found based on a submitted data set with 3 failure points and approximately 30,000 right-censored, suspension, values. After preparation of the contour map, it was felt that this could represent a good test case for a simplistic "Weibayes", or 1-parameter Weibull calculation represented by the source code line to identify the most likely eta from a given beta parameter value:

```
eta_est <- sum((times^beta)/nfail)^(1/beta)
```

Figure 5 plots a series of points from this calculation on a rigorously calculated contour map. It is actually quite stunning to note the proximal agreement of the method with no likelihood calculation.

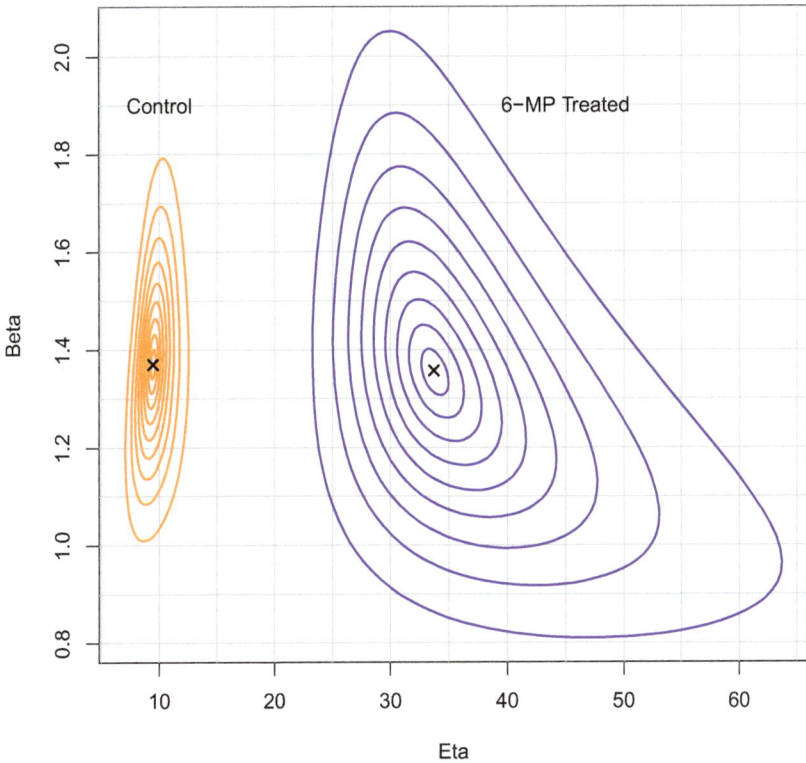

Fig. 3. Comparison of datasets by likelihood contour.

2.2   Likelihood Ratio Bounds

The points on a specific confidence level contour are used to define confidence interval bounds. Figure 5 shows how the extreme Beta value points form asymptotes for the bounds on a 2-parameter model.

2.3   Confidence Interval Bounds on 3-parameter models

Based on requests from users there has been encouragement to implement confidence interval bounds on 3-parameter models. So far all calculated models in `WeibullR` have been easily found in published (Abernethy, 2008; Nelson, 1982; Johnson, 1964; William & Escobar, 1998) examples and some in commercial software. However this has not been observed for cases having modified threshold.

As far as developers of `WeibullR` are concerned, the establishment of likelihood ratio bounds on such 3-parameter models is novel work. The approach used has been to vary the threshold value through a selection of increments from

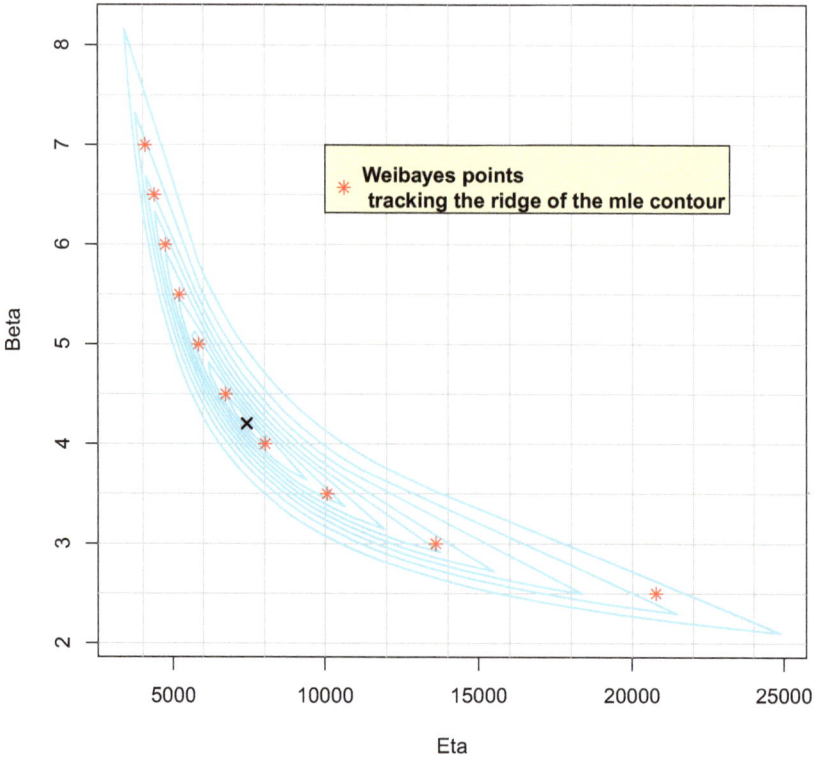

Fig. 4. Testing the Wiebayes function

the optimized MLE. Then 2-parameter contours based on a likelihood ratio determined from the final optimized MLE are plotted as if looking down on a 3 dimensional plot with threshold values in the z-direction. The bounds are then derived from the farthest extent of all the contours.

There is a bound to the maximum upper threshold at the minimum data point. However, in some cases, particularly where the fit may be poor, it has been found that negative threshold values can be unbounded indefinately. In these cases one asumptote is a vertical line. In Figure 6 the same 2-parameter contour and bounds from Figure 5 are displayed as dashed yellow lines, the 3-parameter bounds are the red lines on the modified data plot.

Code and examples (Hensel, 2019; Gelissen, 2017) have been found for the Fisher Matrix bounds including uncertainty in the third parameter. The data used for Figures 5 and 6 have been applied to form these bounds as bold purple lines in Figure 7. The unusual non-convexity in the upper bound has been confirmed using Minitab software.

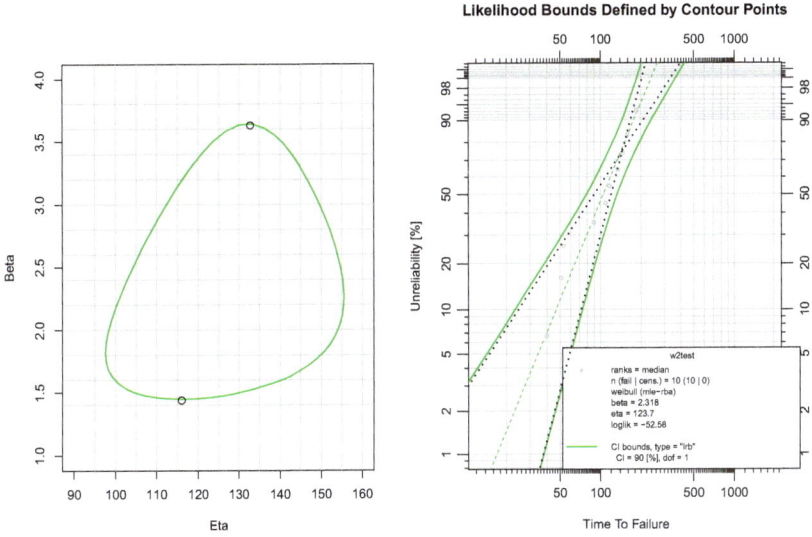

Fig. 5. Likelihood ratio bounds formed by confidence level contour.

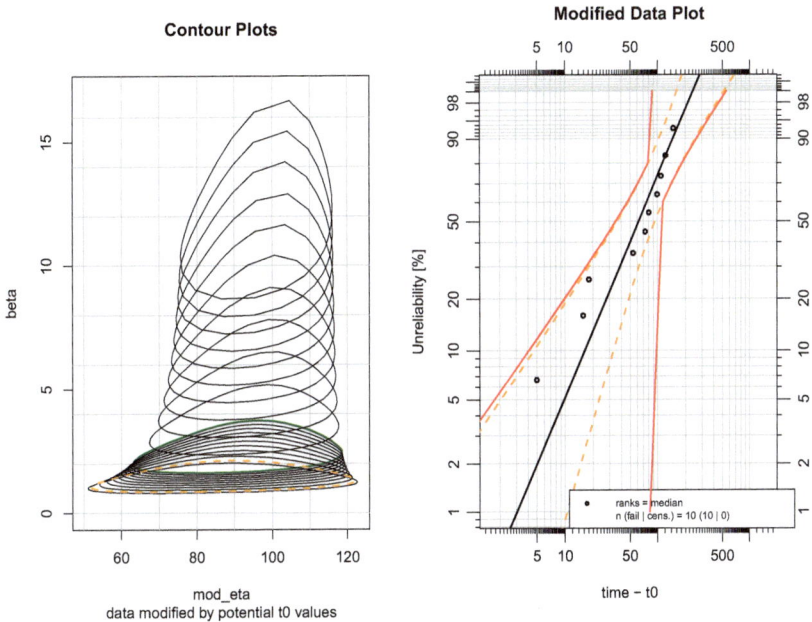

Fig. 6. Contours and bounds for a 3-parameter model.

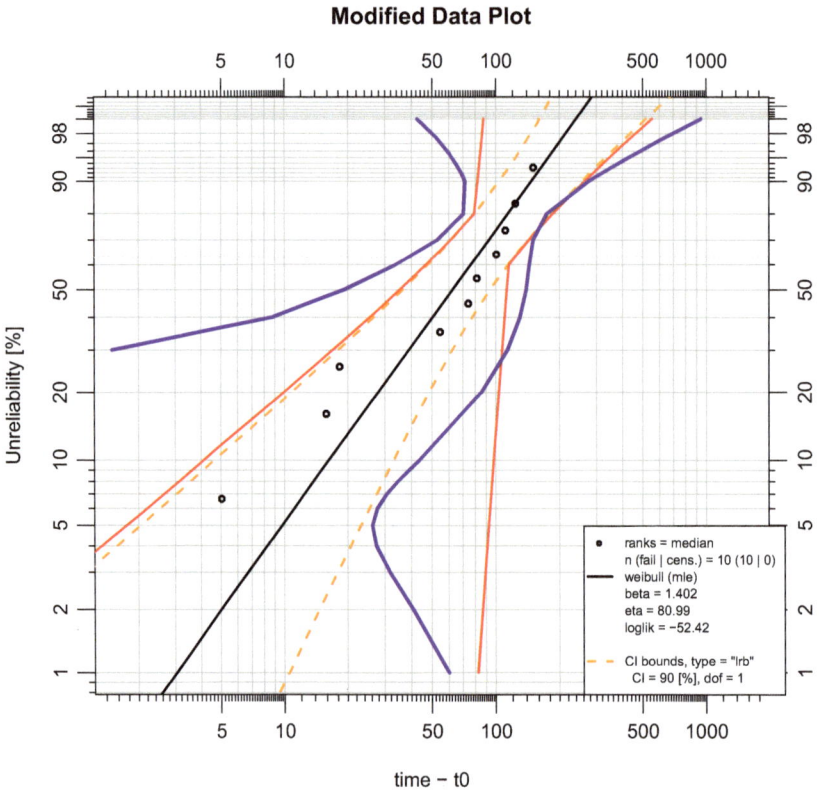

Fig. 7. Unusually formed Fisher Matrix bounds on a 3-parameter model.

3   Grouped Data

3.1   Inspection Data

Grouped inspection data can be handled using an interval data input capability of WeibullR. The latest known working time for a component is the 'left' entry, while the inspection time in which failure was detected is the 'right' entry. It is valid to have a 'left' entry value of zero. These kinds of left-censored data are called discoveries. Suspension, or right-censored, data typically is present for unfailed units at the last inspection time, these must be entered in a 'time' | 'event' | 'qty' dataframe as the primary argument to wblr.

A useful example (Nelson, 1982) with inspection of parts for cracks is reproduced here.

Table 1. Part Cracking Data

Inspection (months)		Number		Cumulative
Start	End	Cracked	Cumulative	
0	6.12	5	5	2.99
6.12	19.92	16	21	12.6
19.92	29.64	12	33	19.8
29.64	35.40	18	51	30.5
35.40	39.72	18	69	41.3
39.72	45.24	2	71	42.5
45.24	52.32	6	77	46.1
52.32	63.48	17	94	56.3
63.48	+ Survived	73	167	

Input data objects for wblr on this data are formed by using start end end times for inspection intervals as follows:

```
inspect <- data.frame(
left=c(0, 6.12, 19.92, 29.64, 35.4, 39.72, 45.32, 52.32),
right=c(6.12, 19.92, 29.64, 35.4, 39.72, 45.32, 52.32, 63.48),
qty=c(5, 16, 12, 18, 18, 2, 6, 17) )
```

The parts surviving at the end of test period are right censored suspensions.

```
survived <- data.frame(time=63.48, event=0, qty=73)
```

The wblr object model uses S3 registration for R's plot method, so a simple call to plot on an object will work. Interval data is represented by horizontal lines across the span of the interval. The mean point of each span is used for placement on the probability plot.

Parts Cracking Inspection Interval Analysis

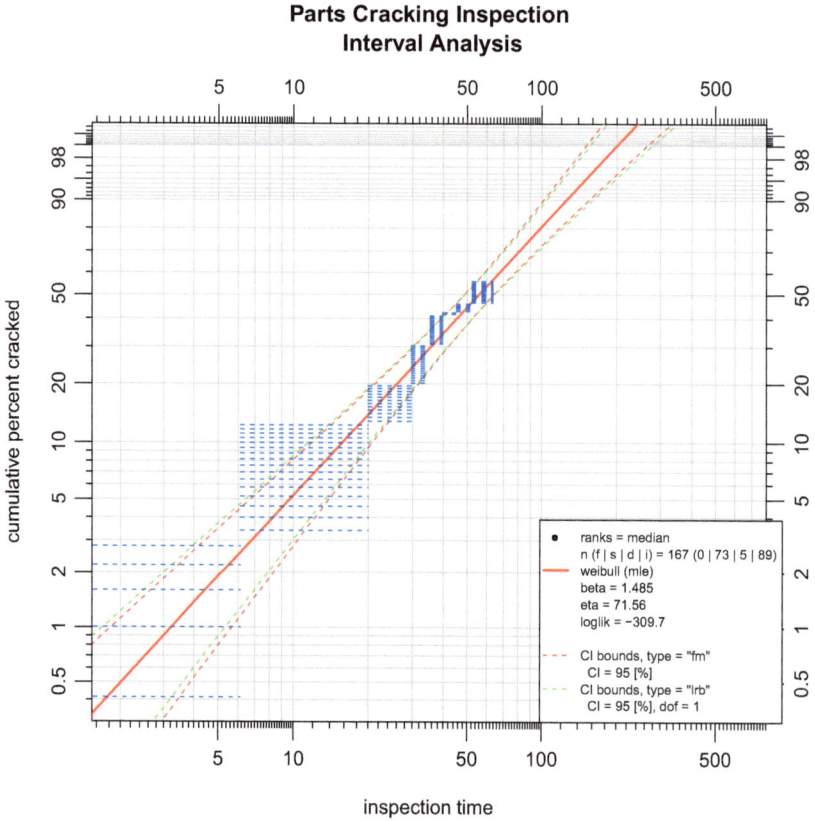

Fig. 8. A probability plot on interval data

3.2 Warranty Data

Warranty data is often analyzed as grouped inspection data. Products placed in service over a period such as a month form a group. Warranty claims are recorded over monthly intervals. As time progresses a layer-cake data form takes shape.

A pending function, `wblr.warranty`, will abstract data wrangling of the values from such a table to form a `wblr` object, from which the user can complete fitting and confidence bound specifications.

Table 2. Layer-Cake Warranty Data

Units Placed in Service Each Month										
97	106	112	116	98	106	108	93	107	101	89
Warranty Claims for Each Group										
9	5	12	7	5	3	6	6	10	9	6
5	4	3	1	6	3	0	3	1	3	3
2	2	1	0	1	6	4	2	2	2	
2	3	4	2	1	2	3	3	1		
4	3	3	1	4	2	1	1			
4	1	2	1	1	1	1				
1	1	2	2	1	1					
2	4	0	4	3						
3	1	1	2							
2	3	1								
2	2									
2										

4 Future Direction

- For the warranty data, the code for data wrangling and analytical display should be encapsulated in easy to use functions.
- Warranty analysis is often associated with Reliability Growth charts utilizing the power law.
- Considerable work is envisioned to provide useful functions to encapsulate considerations to be made with accelerated life testing.
- Several predictive analytics methods are a likely contribution.
- A new package(s) that will depend on `WeibullR`. Likely names may be `WeibullR.applications`, `WeibullR.specialties`, `WeibullR.ALT`, `WeibullR.RGA`.

References

Abernethy, R. (2008). *The new weibull handbook* (5th edition ed.). North Palm Beach, Florida.

Gelissen, S. (2017). *R code for fitting a three-parameter weibull distribution.* ([Blog post] Retrieved from http://blogs2.datall-analyse.nl/2016/02/17/ rcode_three_parameter_weibull/)

Hensel, T. (2019). weibulltools: Statistical methods for life data analysis r package version 1.0.1 [Computer software manual].

Johnson, L. G. (1964). The statistical treatment of fatigue experiments. *ELSEVIER*.

Nelson, W. (1982). *Applied life data analysis.* John Wiley & Sons.

William, W., & Escobar, L. A. (1998). *Statistical methods for reliability data.* John Wiley & Sons Inc.

Pivot Analysis in Weighted Linear Regression

Yuancheng Si

School of Mathematics, University of Manchester, Manchester M13 9PL, UK
yuancheng.si@postgrad.manchester.ac.uk

Abstract. According to Lutzer (2017), the simple linear regression lines based on repeating single observations from a given dataset pivot at a certain pivotal point. In this paper, we discuss this behavior in a more general case and give an explanation about the pivot behavior.

Keywords: Least square regression · Pivot behavior · Weighted least square.

DOI: 10.35566/isdsa2019c4

1 Introduction

In Lutzer (2017), the author investigated how single repeating data point (x_i, y_i) affected the ordinary least squares regression and the result showed that the regression lines pivoted at a certain point under specific condition. Specifically, when $\sum_{j \neq i}^{n} x_j \neq (n-1)x_i$, the pivot point P has coordinate (x_p, y_p) as

$$x_P = x_i + \frac{\sum_{j \neq i}^{n}(x_j - x_i)^2}{\sum_{j \neq i}^{n}(x_j - x_i)} \tag{1}$$

$$y_P = y_i + \frac{\sum_{j \neq i}^{n}(x_j - x_i)(y_j - y_i)}{\sum_{j \neq i}^{n}(x_j - x_i)}. \tag{2}$$

An example is shown in Figure 1. The study background of Lutzer (2017) is based on predicting the result of sport game using match information, in which the explanatory variable and the response variable are both precise positive integer numbers (like rank or number of games) where the occurances of repeated observation always exist. In Lutzer (2017), we found that if certain data point was repeated several times and we employed the simple linear regression based on updated dataset each time, all the simple linear regression lines would intersect at some pivot point P.

However, if the repeated data point, for simplicity we denote it as (x_1, y_1), is not identical with P, the series of squares of residuals after regression for repeated points (x_1, y_1) will converge to 0. The series of squares of residuals for non-repeated data points $\hat{\epsilon_2}^{k^2}, \hat{\epsilon_3}^{k^2} ..., \hat{\epsilon_n}^{k^2}$ converge to a fixed non-zero constant as the number of repeated times k increases, which could lead to heteroscedasticity and

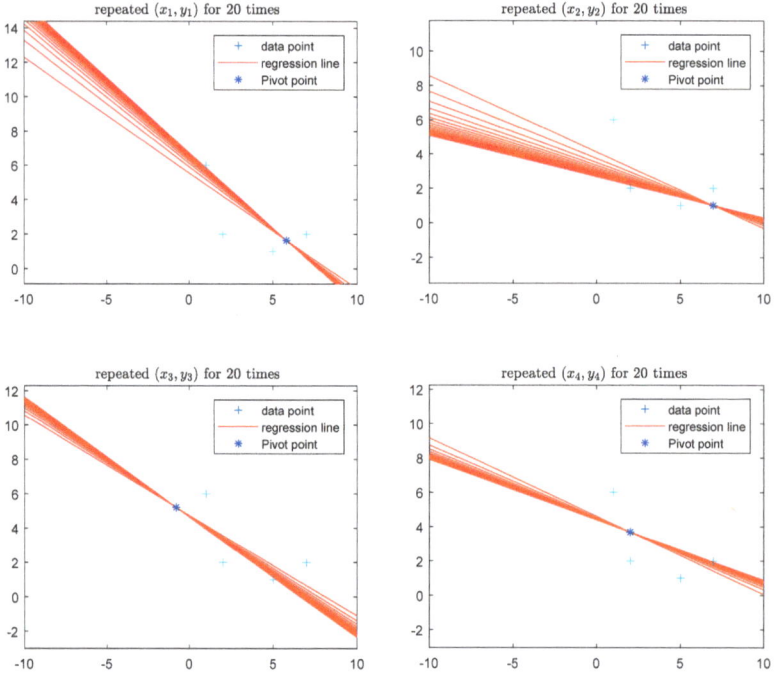

Fig. 1. The pivot properties by repeating each single data point 20 times based on given dataset.

does not satisify the Gauss-Markov theorem. Therefore, the best linear unbiased estimator (BLUE) property of the coefficients given by ordinary least squares regression is not valid any more.

For example, take the dataset given in Lutzer (2017): $(1, 6), (2, 2), (5, 1), (7, 2)$. If we repeat $(2, 2)$ for an additional 19 times and conduct regression using the OLS method, we could find that the simple linear model is no longer valid based on result of either t-test or F-test. We can also exert heteroscedasticity tests using the Breusch-Pagan test or the Whites General Test to detect this issue.

```
Linear regression model:
y ~ 1 + x1

Estimated Coefficients:
              Estimate      SE        tStat      pValue

              --------    -------    -------    ---------

(Intercept)    2.6726     0.36703     7.2817    2.713e-07
x1            -0.23894    0.1426     -1.6756    0.10796
```

Number of observations: 24, Error degrees of freedom: 22
Root Mean Squared Error: 0.819
R-squared: 0.113, Adjusted R-Squared 0.0729
F-statistic vs. constant model: 2.81, p-value = 0.108

    Meanwhile, in real world, different observations have different reliability. For example, we cannot treat the result of friendly match and final match with equal importance. Thus, we should treat different observations with different importances respectively, which inspires us to investigate whether the pivot behavior exists in robust linear models. In this paper, we mainly focus on the pivot behavior based on the weighted least square (WLS) method.

2   The Relationship between Regression Line and Repeated Point

To be more generalized, suppose that we start with the points $(x_1, y_1), (x_2, y_2), ...,$ (x_n, y_n) and we consistently use (x_1, y_1) as the repeated point in the rest of the paper. Note that we repeat (x_1, y_1) for additional k times so we have design matrix $X \in \mathbb{R}^{(n+k) \times 2}$ as

$$
k+1 \text{ times} \left\{ \begin{bmatrix} x_1 & 1 \\ \vdots & \vdots \\ x_1 & 1 \\ x_2 & 1 \\ \vdots & \vdots \\ x_n & 1 \end{bmatrix}_{(n+k) \times 2} \right.
$$

Assume what we have prior information about the weight matrix $W \in \mathbb{R}^{(n+k) \times (n+k)}$ as

$$
\begin{bmatrix} w_{1,1} & \cdots & 0 & 0 & \cdots & 0 \\ \vdots & \ddots & \vdots & \vdots & \ddots & \vdots \\ 0 & \cdots & w_{1,k+1} & 0 & \cdots & 0 \\ 0 & \cdots & 0 & w_2 & \cdots & 0 \\ \vdots & \ddots & \vdots & \vdots & \ddots & \vdots \\ 0 & \cdots & 0 & 0 & \cdots & w_n \end{bmatrix}_{(n+k) \times (n+k)}
$$

where $w_{1,1}, ..., w_{1,k+1} \in \mathbb{R}^+$ are the corresponding weights for a total of $k + 1$ repeated points (x_1, y_1), respectively, and $w_2, ..., w_n \in \mathbb{R}^+$ are weights for non-repeated points. Note that we did not discuss how to determine the optimal weight scheme for each observation in this article. It depends on the personal judgement on the precision of measurment or specific weight scheme that could make the estimator of unknown parameters significant under the WLS method. Furthermore, the weights $w_{1,1}, ..., w_{1,k+1}$ for a total of $k + 1$ times of repeated data points (x_1, y_1) do not need to be identical and their sum do not need to add up to 1 under our parameterization.

Suppose that the model under consideration is

$$Y = X\beta + \epsilon \tag{3}$$

where

$$\beta = \begin{bmatrix} m_k \\ b_k \end{bmatrix} \tag{4}$$

$$\epsilon \sim N(0, \sigma^2 W) \tag{5}$$

$$Y = \begin{bmatrix} y_1 \\ \vdots \\ y_1 \\ y_2 \\ \vdots \\ y_n \end{bmatrix} \left.\begin{array}{}\\\\\\\end{array}\right\} k+1 \text{ times} \quad .$$

The normal equation

$$X^T W X \beta = X^T W Y \tag{6}$$

becomes

$$\begin{bmatrix} x_1^2 \sum_{i=1}^{k+1} w_{1,i} + \sum_{i=2}^{n} w_i x_i^2, & x_1 \sum_{i=1}^{k+1} w_{1,i} + \sum_{i=2}^{n} w_i x_i \\ x_1 \sum_{i=1}^{k+1} w_{1,i} + \sum_{i=2}^{n} w_i x_i, & \sum_{i=1}^{k+1} w_{1i} + \sum_{i=2}^{n} w_i \end{bmatrix} \begin{bmatrix} m_k \\ b_k \end{bmatrix} \tag{7}$$

$$= \begin{bmatrix} x_1 y_1 \sum_{i=1}^{k+1} w_{1,i} + \sum_{i=2}^{n} w_i x_i y_i \\ y_1 \sum_{i=1}^{k+1} w_{1,i} + \sum_{i=2}^{n} w_i y_i \end{bmatrix}. \tag{8}$$

From the second component of the normal equation, we have

$$\frac{x_1 \sum_{i=1}^{k+1} w_{1,i} + \sum_{i=2}^{n} w_i x_i}{\sum_{i=1}^{k+1} w_{1,i} + \sum_{i=2}^{n} w_i} m_k + b_k = \frac{y_1 \sum_{i=1}^{k+1} w_{1,i} + \sum_{i=2}^{n} w_i y_i}{\sum_{i=1}^{k+1} w_{1,i} + \sum_{i=2}^{n} w_i}, \tag{9}$$

which means that each regression line includes weighted average means of the $x-$ and $y-$ coordinates of the total $n+k$ data points. We denote it as $M_k(\mu_{x;k}, \mu_{y;k})$ where

$$\mu_{x;k} = \frac{x_1 \sum_{i=1}^{k+1} w_{1,i} + \sum_{i=2}^{n} w_i x_i}{\sum_{i=1}^{k+1} w_{1,i} + \sum_{i=2}^{n} w_i} \tag{10}$$

$$\mu_{y;k} = \frac{y_1 \sum_{i=1}^{k+1} w_{1,i} + \sum_{i=2}^{n} w_i y_i}{\sum_{i=1}^{k+1} w_{1,i} + \sum_{i=2}^{n} w_i}. \tag{11}$$

$(\mu_{x;k}, \mu_{y;k})$ becomes $(\bar{x}, \bar{y})$ if we assign equal weight for each observation, i.e., $w_{1,1} = \cdots = w_{1,k+1} = w_2 = \cdots = w_n$. Therefore, the ordinary least square regression line must go through the mean point of the data.

Furthermore, the limiting position of $\mu_{x;k}$ and $\mu_{y;k}$, as $k \to +\infty$, depends on the convergence of $\sum_{i=1}^{k+1} w_{1,i}$. That is, if $\lim_{k \to +\infty} \sum_{i=1}^{k+1} w_{1,i} = +\infty$ we have

$$\lim_{k \to +\infty} \mu_{x;k} = x_1 \tag{12}$$

$$\lim_{k \to +\infty} \mu_{y;k} = y_1, \tag{13}$$

i.e., $M_k \to P(x_1, y_1)$ as k grows.

If $\lim_{k \to +\infty} \sum_{i=1}^{k+1} w_{1,i} = c \in \mathbb{R} \geq 0$ we have

$$\lim_{k \to +\infty} \mu_{x;k} = \frac{x_1 c + \sum_{i=2}^{n} w_i x_i}{c + \sum_{i=2}^{n} w_i} \tag{14}$$

$$\lim_{k \to +\infty} \mu_{y;k} = \frac{y_1 c + \sum_{i=2}^{n} w_i y_i}{c + \sum_{i=2}^{n} w_i} \tag{15}$$

which does not approach the repeated point but a new point R_{skew}.

Note that the slope of the line segment connecting M_k and M_{k+1} is given by

$$\frac{\mu_{y;(k+1)} - \mu_{y;k}}{\mu_{x;(k+1)} - \mu_{x;k}} = \frac{y_1(\sum_{i=2}^{n} w_i) - \sum_{i=2}^{n} w_i y_i}{x_1(\sum_{i=2}^{n} w_i) - \sum_{i=2}^{n} w_i x_i}. \tag{16}$$

If Eq (16) is invariant for different choices of k, we can conclude that the weighted average points of the data is collinear because the quotient is independent of k. In fact, it mainly relies on our weight scheme setup for the updated data. For example, if we assume that newly added repeated point (x_1, y_1) will not influence the weights of non-repeated points $w_2, ..., w_n$, Eq (16) will be independent of k. In contrast, if we assume that the newly added repeated point will increase/decrease the original weights for non-repeated points in a linear pattern, for example, $w_i = w_i + c_i k$, the collinear property for weighted average points $M_0, M_1, ..., M_k$ could disappear for different selections of c_i.

3   Finding the Pivot Point

Based on the method we used above, we tried to apply it to the first component of the Eq (7) to find the pivot behavior. When $x_1 \sum_{i=1}^{k+1} w_{1,i} + \sum_{i=2}^{n} w_i x_i \neq 0$, we have

$$\frac{x_1^2 \sum_{i=1}^{k+1} w_{1,i} + \sum_{i=2}^{n} w_i x_i^2}{x_1 \sum_{i=1}^{k+1} w_{1,i} + \sum_{i=2}^{n} w_i x_i} m + b = \frac{x_1 y_1 \sum_{i=1}^{k+1} w_{1,i} + \sum_{i=2}^{n} w_i x_i y_i}{x_1 \sum_{i=1}^{k+1} w_{1,i} + \sum_{i=2}^{n} w_i x_i}. \tag{17}$$

Using the coordinate system exchange from Cartesian coordinate to R-centric coordinate ($\chi = x - x_1, \gamma = y - y_1$), we express Eq (17) as

$$\frac{\sum_{i=2}^{n} w_i \chi_i^2}{\sum_{i=2}^{n} w_i \chi_i} m_k + b_k = \frac{\sum_{i=2}^{n} w_i \chi_i \gamma_i}{\sum_{i=2}^{n} w_i \chi_i}. \tag{18}$$

From Eq (18), we clearly see that for any choice of k, all the regression lines will intersect at point P under the R-centric coordinate with the proper weight scheme

$$\left(\frac{\sum_{i=2}^{n} w_i \chi_i^2}{\sum_{i=2}^{n} w_i \chi_i}, \frac{\sum_{i=2}^{n} w_i \chi_i \gamma_i}{\sum_{i=2}^{n} w_i \chi_i} \right) \tag{19}$$

or

$$\left(x_1 + \frac{\sum_{i=2}^{n} w_i (x_i - x_1)^2}{\sum_{i=2}^{n} w_i (x_i - x_1)}, y_1 + \frac{\sum_{i=2}^{n} w_i (x_i - x_1)(y_i - y_1)}{\sum_{i=2}^{n} w_i (x_i - x_1)} \right) \tag{20}$$

under the Cartesian coordinate.

We should notice that the regression lines pivot at P with two prerequiste conditions.

1. $\sum_{i=2}^{n} w_i \chi_i \neq 0$ under the R-centric coordiante. If $\sum_{i=2}^{n} w_i \chi_i = 0$, Eq (7) could be written as

$$\begin{bmatrix} \sum_{i=2}^{n} w_i \chi_i^2, & 0 \\ 0, & \sum_{i=1}^{k+1} w_{1,i} + \sum_{i=2}^{n} w_i \end{bmatrix} \begin{bmatrix} m_k \\ b_k \end{bmatrix} = \begin{bmatrix} \sum_{i=2}^{n} w_i \chi_i \gamma_i \\ \sum_{i=2}^{n} w_i \gamma_i \end{bmatrix}. \tag{21}$$

where we have a cloud of parallel regression lines with the identical slope

$$m_k = \frac{\sum_{i=2}^{n} w_i \chi_i \gamma_i}{\sum_{i=2}^{n} w_i \chi_i^2}$$

but different intercepts

$$b_k = \frac{\sum_{i=2}^{n} w_i \gamma_i}{\sum_{i=1}^{k+1} w_{1,i} + \sum_{i=2}^{n} w_i}.$$

We can see that b is not invariant from k.

The figure below (Figure 2) depicts the case when $\sum_{i=2}^{n} w_i \chi_i = 0$ holds for all equal weights for repeated and non-repeated data points. We amend the data as $(6,6), (2,2), (5,1), (7,2)$. If we repeat the third data point $(5,1)$, as it satisties the condition $6 + 2 + 7 = 5 * (4 - 1)$, we have a set of parallel regression lines without even a single intersection between any two regression line.

2. The weight scheme plays a latent but vital role in determining whether the pivot point exist. Actually, even if the newly added repeated point affects the weights of non-repeated data points, provided that Eq (20) keeps invariant, we can always find a pivot behavior. For example, if $w_i \propto k$ or $w_i = c_i/k$ for $i = 2, ..., n$, we will still find a pivot point for the regression lines.

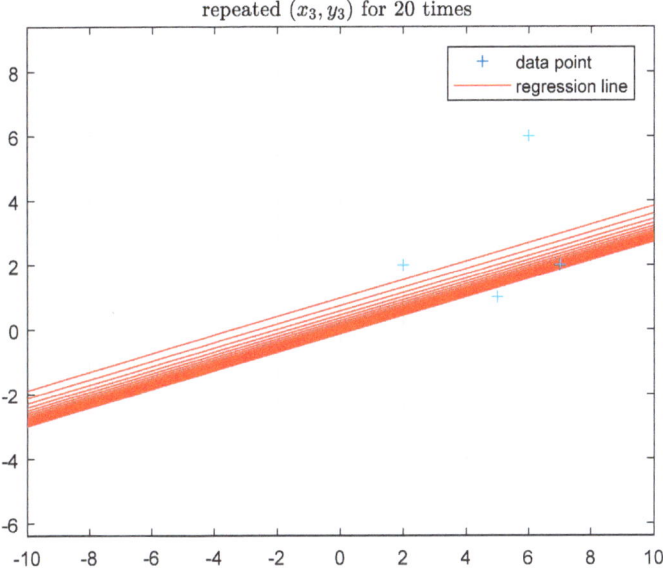

Fig. 2. Repeat $(5,1)$ in the data $(6,6),(2,2),(5,1),(7,2)$

4   Further Discussion on the Pivot Point

For the origin of the pivot point, in order to simplify our calculation, we express the following calculation in R-centric coordinate. The normal equation under the R-centric coordinate system is given below:

$$\begin{bmatrix} \sum_{i=2}^n w_i\chi_i^2, & \sum_{i=2}^n w_i\chi_i \\ \sum_{i=2}^n w_i\chi_i, & \sum_{i=1}^{k+1} w_{1,i} + \sum_{i=2}^n w_i \end{bmatrix} \begin{bmatrix} m_k \\ b_k \end{bmatrix} = \begin{bmatrix} \sum_{i=2}^n w_i\chi_i\gamma_i \\ \sum_{i=2}^n w_i\gamma_i \end{bmatrix}. \qquad (22)$$

If the left hand side matrix $X^T W X$ is non-singular, we could solve this normal equation with a specific repeated time k

$$m_k = \frac{(\sum_{i=1}^{k+1} w_{1,i} + \sum_{i=2}^n w_i)(\sum_{i=2}^n \chi_i\gamma_i w_i) - (\sum_{i=2}^n \chi_i w_i)(\sum_{i=2}^n \gamma_i w_i)}{(\sum_{i=2}^n \chi_i^2 w_i)(\sum_{i=1}^{k+1} w_{1,i} + \sum_{i=2}^n w_i) - (\sum_{i=2}^n \chi_i w_i)^2}$$
$$b_k = \frac{(\sum_{i=2}^n \chi_i^2 w_i)(\sum_{i=2}^n \gamma_i w_i) - (\sum_{i=2}^n \chi_i w_i)(\sum_{i=2}^n \chi_i\gamma_i w_i)}{(\sum_{i=2}^n \chi_i^2 w_i)(\sum_{i=1}^{k+1} w_{1,i} + \sum_{i=2}^n w_i) - (\sum_{i=2}^n \chi_i w_i)^2} \qquad (23)$$

In order to control variations between weights, we assume that the newly added repeated points will not affect the weights of the rest of the $n+k-1$ data points, i.e., $w_{1,1}, ..., w_{1,k}, w_2, ..., w_n$ keep the same when $w_{1,k+1}$ is added.

4.1   Results under Some Certain Weight Schemes

We consider the following results as we apply different weight scheme to the data.

1. When $\lim\limits_{k \to +\infty} \sum_{i=1}^{k+1} w_{1,i} = +\infty$, we have

$$m = \lim_{k \to +\infty} m_k = \frac{\sum_{i=2}^{n} \chi_i \gamma_i w_i}{\sum_{i=2}^{n} \chi_i^2 w_i}, \tag{24}$$

$$b = \lim_{k \to +\infty} b_k = 0. \tag{25}$$

Based on this result and Eq (12), we can conclude that the limit position of the regression line must go through the origin $(0,0)$ and the pivot point (20). Actually, we can calculate the slope of limit position of a regression line directly through (20)

$$\frac{\sum_{i=2}^{n} w_i \chi_i \gamma_i}{\sum_{i=2}^{n} w_i \chi_i} \Big/ \frac{\sum_{i=2}^{n} w_i \chi_i^2}{\sum_{i=2}^{n} w_i \chi_i}$$

which is in accord with Eq (24).

2. If $\lim\limits_{k \to +\infty} \sum_{i=1}^{k+1} w_{1,i} = c \in \mathbb{R}$ where $c \geq 0$,

$$m = \lim_{k \to +\infty} m_k = \frac{(c + \sum_{i=2}^{n} w_i)(\sum_{i=2}^{n} \chi_i \gamma_i w_i) - (\sum_{i=2}^{n} \chi_i w_i)(\sum_{i=2}^{n} \gamma_i w_i)}{(\sum_{i=2}^{n} \chi_i^2 w_i)(c + \sum_{i=2}^{n} w_i) - (\sum_{i=2}^{n} \chi_i w_i)^2},$$

$$b = \lim_{k \to +\infty} b_k = \frac{(\sum_{i=2}^{n} \chi_i^2 w_i)(\sum_{i=2}^{n} \gamma_i w_i) - (\sum_{i=2}^{n} \chi_i \gamma_i w_i)(\sum_{i=2}^{n} \chi_i w_i)}{(\sum_{i=2}^{n} \chi_i^2 w_i)(c + \sum_{i=2}^{n} w_i) - (\sum_{i=2}^{n} \chi_i w_i)^2}. \tag{26}$$

We can also use the coordinate of R_{skew} (Eq (14) converted to R-centric coordinate system first) and the pivot point P to calculate the slope of the limit position of a regression line.

The figure below (Figure 3) shows that we assign the weight to the k-th newly added (x_1, y_1) (in total repeated 100 times) with weight $w_{1,k} = \frac{1}{2^k}$ where the remaining non-repeated points $(x_2, y_2), (x_3, y_3), (x_4, y_4)$ with invariant weights $w_2 = 2, w_3 = 3, w_4 = 4$ respectively. We can still find a pivot point for regression lines but it will no longer approach to (x_1, y_1) no matter how many times the point is repeated. In contrast, in Figure 1, (x_1, y_1) was only repeated for 20 times.   From the figure, we notice that the limit position of the regression line could not go through the repeated point R as $k \to +\infty$ (or the limit line may not go through the origin $(0,0)$ in the R-centric coordinate system.) because the sum of weights of the repeated point is a finite constant $c \in \mathbb{R}$ as $k \to +\infty$.

3. Each weight is identical and free from k, i.e., $w_{1,1} = w_{1,2} = ... = w_{1,(k+1)} = w_2 = ... = w_n = w \in \mathbb{R}^+$, which can be viewed as OLS regression and it is a

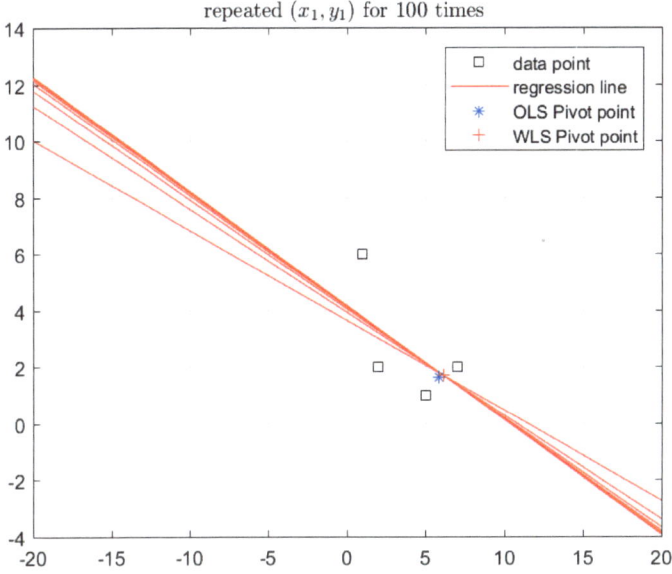

Fig. 3. Repeat $(1, 6)$ in $(1, 6), (2, 2), (5, 1), (7, 2)$ for 100 times

special case for case 1, we have

$$m = \lim_{k \to +\infty} m_k = \frac{\sum_{i=2}^{n} \chi_i \gamma_i}{\sum_{i=2}^{n} \chi_i^2} \tag{27}$$

$$b = \lim_{k \to +\infty} b_k = 0 \tag{28}$$

4.2   Explanations of the Pivot Point

We denote $\tilde{l} = mx + b$ as the limit position of the regression lines. I am going to show that for any given repeating time k, the line $\gamma_k = m_k \chi + b_k$ is a weighted average of $l_k : \tilde{l}$ and l_0 with weight w_k, which is equivalent to prove:

$$\begin{aligned} b_k &= w_k b_0 + (1 - w_k) b \\ m_k &= w_k m_0 + (1 - w_k) m \end{aligned} \tag{29}$$

where

$$m_0 = \frac{(w_{1,1} + \sum_{i=2}^{n} w_i)(\sum_{i=2}^{n} \chi_i \gamma_i w_i) - (\sum_{i=2}^{n} \chi_i w_i)(\sum_{i=2}^{n} \gamma_i w_i)}{(\sum_{i=2}^{n} \chi_i^2 w_i)(w_{1,1} + \sum_{i=2}^{n} w_i) - (\sum_{i=2}^{n} \chi_i w_i)^2} \tag{30}$$

$$b_0 = \frac{(\sum_{i=2}^{n} \chi_i^2 w_i)(\sum_{i=2}^{n} \gamma_i w_i) - (\sum_{i=2}^{n} \chi_i \gamma_i w_i)(\sum_{i=2}^{n} \chi_i w_i)}{(\sum_{i=2}^{n} \chi_i^2 w_i)(w_{1,1} + \sum_{i=2}^{n} w_i) - (\sum_{i=2}^{n} \chi_i w_i)^2}. \tag{31}$$

We have obtained the expression for m_k and b_k from Eq (23) and m and b from Eq (26) and (23).

It is clear that

$$w_k = \frac{(\sum_{i=2}^n \chi_i^2 w_i)(w_{1,1} + \sum_{i=2}^n w_i) - (\sum_{i=2}^n \chi_i w_i)^2}{(\sum_{i=2}^n \chi_i^2 w_i)(\sum_{i=1}^{k+1} w_{1,i} + \sum_{i=2}^n w_i) - (\sum_{i=2}^n \chi_i w_i)^2} \tag{32}$$

under Case 1 in the subsection 4.1.

$$w_k = \frac{(\sum_{i=2}^n \chi_i^2 w_i)(c + \sum_{i=2}^n w_i) - (\sum_{i=2}^n \chi_i w_i)^2}{(\sum_{i=2}^n \chi_i^2 w_i)(\sum_{i=1}^{k+1} w_{1,i} + \sum_{i=2}^n w_i) - (\sum_{i=2}^n \chi_i v_i)^2} \tag{33}$$

under Case 2 in the subsection 4.1.

That is, each l_k is a weighted average of l_0 and $\tilde{l}$. So the intersection between l_0 and $\tilde{l}$ must appear on each weighted line between l_0 and $\tilde{l}$, which is exactly the pivot point we find.

5   An Extension to the Simulation of Pivot Behavior in 3-Dimension

In this section, I will try to find the pivot behavior in the three dimensional data cases under the equal weight scheme. Suppose that we want to pass a plane $y = k_1 x_1 + k_2 x_2 + b$ through the points $\underbrace{(2, -1, 4)...(2, -1, 4)}_{k+1 \text{ times}}, (-1, 3, -2), (0, 2, 3),$
and $(-1, -2, 0)$ using the OLS method. We have a total of $k + 4$ data points, $k+1$ of them are $(2, -1, 4)$ with 1 original point and k repetitions. We can select any single point to repeat. For simplicity, we repeat $(x_{11}, x_{21}, y_1) = (2, -1, 4)$. We centralize the whole datasets with respect to the repeated point by setting $(x_{11}, x_{21}, y_1) = (2, -1, 4)$ as original point $(0, 0, 0)$, shifting in space, so the dataset turns out to be $\underbrace{(0, 0, 0)...(0, 0, 0)}_{k+1 \text{ times}}, (-3, 4, -6), (-2, 3, -1),$ and $(-3, -1, -4)$.

More generally, a point with the Cartesian coorinates (x_{i1}, x_{i2}, y_i) has the R−centric coordinates $\chi_{i1} = x_{i1} - x_{11}, \chi_{i2} = x_{i2} - x_{12}, v_i = y_i - y_1$. Fitting the plane $v = k_1 \chi_1 + k_2 \chi_2 + b$ to the data is equivalent to

$$k+1 \text{ times} \left\{ \begin{bmatrix} 0 & 0 & 1 \\ \vdots & \vdots & \vdots \\ 0 & 0 & 1 \\ -3 & 4 & 1 \\ -2 & 3 & 1 \\ -3 & -1 & 1 \end{bmatrix} \begin{bmatrix} k_1 \\ k_2 \\ b \end{bmatrix} + \epsilon = \begin{bmatrix} 0 \\ \vdots \\ 0 \\ -6 \\ -1 \\ -4 \end{bmatrix} \right.$$

with the associated normal equation

$$\begin{bmatrix} \sum_{i=2}^4 \chi_{i1}^2, & \sum_{i=2}^4 \chi_{i1}\chi_{i2}, & \sum_{i=2}^4 \chi_{i1} \\ \sum_{i=2}^4 \chi_{i1}\chi_{i2}, & \sum_{i=2}^4 \chi_{i2}^2, & \sum_{i=2}^4 \chi_{i2} \\ \sum_{i=2}^4 \chi_{i1}\chi_{i1}, & \sum_{i=2}^4 \chi_{i1}\chi_{i2}, & k+n \end{bmatrix} \begin{bmatrix} k_1 \\ k_2 \\ b \end{bmatrix} = \begin{bmatrix} \sum_{i=2}^4 \chi_{i1}v_i \\ \sum_{i=2}^4 \chi_{i2}v_i \\ \sum_{i=2}^4 v_i \end{bmatrix} .$$

By comparing the first 2 equations from the matrix equation above, we have

$$\sum_{i=2}^{4}\chi_{i1}^2 k_1 + \sum_{i=2}^{4}\chi_{i1}\chi_{i2}k_2 + \sum_{i=2}^{4}\chi_{i1}b = \sum_{i=2}^{4}\chi_{i1}v_i \tag{34}$$

$$\sum_{i=2}^{4}\chi_{i1}\chi_{i2}k_1 + \sum_{i=2}^{4}\chi_{i2}^2 k_2 + \sum_{i=2}^{4}\chi_{i2}b = \sum_{i=2}^{4}\chi_{i2}v_i. \tag{35}$$

When $\sum_{i=2}^{4}\chi_{i1} \neq 0$ and $\sum_{i=2}^{4}\chi_{i2} \neq 0$, we can divide them to find the corresponding points on the plane $v = k_1\chi_1 + k_2\chi_2 + b$, namely

$$(\frac{\sum_{i=2}^{4}\chi_{i1}^2}{\sum_{i=2}^{4}\chi_{i1}}, \frac{\sum_{i=2}^{4}\chi_{i1}\chi_{i2}}{\sum_{i=2}^{4}\chi_{i1}}, \frac{\sum_{i=2}^{4}\chi_{i1}v_i}{\sum_{i=2}^{4}\chi_{i1}})$$
$$(\frac{\sum_{i=2}^{4}\chi_{i1}\chi_{i2}}{\sum_{i=2}^{4}\chi_{i2}}, \frac{\sum_{i=2}^{4}\chi_{i2}^2}{\sum_{i=2}^{4}\chi_{i2}}, \frac{\sum_{i=2}^{4}\chi_{i2}v_i}{\sum_{i=2}^{4}\chi_{i2}}). \tag{36}$$

Notice that Eq (36) is not related to k, which means that no matter how many times we repeated, the fitting plane should contain points in Eq (36). If Eq (36) is not identical, we can easily find the pivot line goes through the point in Eq (36), which is given by the symmetric equations of the line:

$$\frac{x_1 - \frac{\sum_{i=2}^{4}\chi_{i1}^2}{\sum_{i=2}^{4}\chi_{i1}}}{\frac{\sum_{i=2}^{4}\chi_{i1}^2}{\sum_{i=2}^{4}\chi_{i1}} - \frac{\sum_{i=2}^{4}\chi_{i1}\chi_{i2}}{\sum_{i=2}^{4}\chi_{i2}}}$$
$$= \frac{x_2 - \frac{\sum_{i=2}^{4}\chi_{i1}\chi_{i2}}{\sum_{i=2}^{4}\chi_{i1}}}{\frac{\sum_{i=2}^{4}\chi_{i1}\chi_{i2}}{\sum_{i=2}^{4}\chi_{i1}} - \frac{\sum_{i=2}^{4}\chi_{i2}^2}{\sum_{i=2}^{4}\chi_{i2}}}$$
$$= \frac{y - \frac{\sum_{i=2}^{4}\chi_{i1}v_i}{\sum_{i=2}^{4}\chi_{i1}}}{\frac{\sum_{i=2}^{4}\chi_{i1}v_i}{\sum_{i=2}^{4}\chi_{i1}} - \frac{\sum_{i=2}^{4}\chi_{i2}v_i}{\sum_{i=2}^{4}\chi_{i2}}}. \tag{37}$$

6   Conclusion

In this article, we studied the pivot behavior in regression using the weighted least square method and gave a plausible explanation for it. However, the weight scheme plays a vital role through the whole study procedure. The dependency between the weight scheme and the pivot behavior could be studied in the future. Meanwhile, the explanation of the pivot behavior could be applied in different scientific areas to gain further insight of the impact coming from repeated data points under the WLS method.

References

Lutzer, C. V. (2017). A curious feature of regression. *The College Mathematics Journal*, *48*(3), 189–198.

MCMC Bootstrap Based Approach to Power and Sample Size Evaluation

Oleksandr Mykolayovich Ocheredko[0000−0002−4792−8581]

National Pirogov Memorial Medical University, 21008 Vinnytsya, Ukraine
https://www.vnmu.edu.ua/en/department/department/10#
Ocheredko@vnmu.edu.ua

Abstract. Power calculation is an important and evergreen applied statistical avenue. This study delivers suggestions on enrichment of the statistical tools by a combination of bootstrap and MCMC modeling. Novelty suggests application of possible data generation mechanism using MCMC and power estimation in the bootstrap procedure. We delineated further generalizations not incorporated in statistical software yet and demonstrated basic applications using `SAS/STAT POWER Procedure` examples (SAS Institute Inc., 2004). One concerns ANOVA, and the other deals with the survival process. The third example deals with the log-linear modeling of thromboembolism data (Congdon, 2005). An illustrious advantage of using MCMC is the possibility to exploit distributions of parameters of interest instead of ubiquitously used point estimates. The other methodological advancement though not demonstrated in the paper is the possibility to combine preliminary or historically observed data with experts' views. The foremost appealing advantage to application environment is the flexibility that is not confined to several basic situations rendered by statistical software. We have chosen the `BUGS` language to demonstrate the program code that can be run on `WinBUGS`, `OpenBUGS` and `JAGS` engines (Lunn, Jackson, Best, Spiegelhalter, & Thomas, 2012). We have used the `R2WinBUGS` package (Gelman., 2015) to run the script with `n.chains=1`, `n.iter=5000`, `bugs.seed=1966` specification.

Keywords: Power Analysis · Data Generation Mechanism · MCMC · Bootstrap.

DOI: 10.35566/isdsa2019c5

1 Introduction

Almost everything in Data Analysis is about hypotheses testing. It is surely not enough to rely on p-value of test. What matters? One has to consider (i) design of data collection, (ii) test appropriateness, and (iii) sample size. Only linking together data generation mechanism, test, and driven sample size, we assuredly safeguard hypotheses testing. Where are we standing now? The ubiquitous approach is to relate alpha and beta errors with sample size given test distribution, resulting in complex formula. As a result, it is stiff to support modern design advancements, such as complex designs with numerous operative units and structured dependencies. The core of the paper is the incorporation of data generation mechanism in sample size evaluation empowered by MCMC that procures:

- flexibility no more confined to several basic situations rendered by statistical software,
- correspondence of power evaluation to the appropriate (say, that incorporates relevant elements of information matrix) test,
- possibility to combine preliminary, historical data, experts experiences with sample, and
- interval against point estimates.

The essentials of approach include:

- implementation of data generation mechanism driven by hypothesis and empowered by MCMC through likelihood function (describes preliminary data, if any, and information from hypothesis) and by priors (describe historical data, assumptions),
- implementation of power levels in the bootstrap procedure.

To demonstrate the practicalities of the approach, I opted for basic applications using SAS/STAT POWER Procedure examples (SAS Institute Inc., 2004). The first example concerns ANOVA, and the second treats survival process. The third example deals with the log-linear modeling of thromboembolism data (Congdon, 2005). I have chosen BUGS language to demonstrate the program code that can be run on WinBUGS, OpenBUGS, JAGS engines (Lunn et al., 2012). I have used R2WinBUGS package (Gelman., 2015) to run the script with n.chains=1, n.iter=5000, bugs.seed =1966 specifications.

Throughout the examples, I tried to demonstrate the flexibility of that approach that is not confined to several basic situations rendered by statistical software. This in my opinion is the foremost appealing feature in applications.

2   Examples

2.1   Example 1. One-Way ANOVA

This example taken from SAS/STAT 9.1 User's Guide (2004), Chapter 57. The POWER Procedure. Example 57.1. One-Way ANOVA (p. 3536) (SAS Institute Inc., 2004).

Hocking (1985, p. 109) describes a study of the effectiveness of electrolytes in reducing lactic acid buildup for long-distance runners. A similar study is planned in which five different fluids will be allocated to runners on a 10-mile course and lactic acid buildup measurements will be taken immediately after the race. The fluids consist of water and two commercial electrolyte drinks, EZDure and LactoZap, each prepared at two concentrations, low (EZD1 and LZ1) and high (EZD2 and LZ2).

Table 1. Mean Lactic Acid Buildup by Fluid

Water	EZD1	EZD2	LZ1	LZ2
35.6	33.7	30.2	29	25.9

The standard deviation of lactic acid measurements given any particular fluid is about 3.75, and the expected lactic acid values correspond to those in Table 1.

Four different comparisons are assumed, shown in Table 2 with appropriate contrast coefficients.

Table 2. Planned Comparisons

Comparison	Contrast Coefficients				
	Water	EZD1	EZD2	LZ1	LZ2
Water versus electrolytes	4	-1	-1	-1	-1
EZD versus LZ	0	1	1	-1	-1
EZD1 versus EZD2	0	1	-1	0	0
LZ1 versus LZ2	0	0	0	1	-1

For each of these contrasts, I estimate the sample size required to achieve a power of 0.9 with the significance level $\alpha = 0.025$. For the interests of reducing costs, a sample size weighting scheme of 2:1:1:1:1 is applied. The **SAS** statements required to perform this analysis are as follows:

```
proc power;
onewayanova
groupmeans = 35.6 | 33.7 | 30.2 | 29 | 25.9
stddev = 3.75
groupweights = (2 1 1 1 1)
alpha = 0.025
ntotal = .
power = 0.9
contrast=(4 -1 -1 -1 -1) (0 1 1 -1 -1)
(0 1 -1 0 0) (0 0 0 1 -1);
run;
```

The output is displayed in Table 3.

Proposed realization Let us consider the first contrast. Others follow the suit. For the generation of means, I suggest the normal means sampled from $Y_i \sim \text{Normal}(\mu_i, \sigma^2/n_i)$. The **BUGS** code for the analysis is given below.

```
model{
    N~dcat(p[])
    for (i in 1:100) { p[i] <- 1/100 }
```

Table 3. SAS output, One-Way ANOVA example

Index	Contrast	Power	NTotal
1	4-1-1-1-1	0.947	30
2	4-1-1-1-1	0.901	24
3	0 1 1-1-1	0.929	60
4	0 1 1-1-1	0.922	48
5	0 1-1 0 0	0.901	174
6	0 1-1 0 0	0.901	174
7	0 0 0 1-1	0.902	222
8	0 0 0 1-1	0.902	480

```
for (i in 2:5) { n[i] <- N }
n[1] <- 2*N
for (i in 1:5) {
    for (k in 1:K) { y[i,k]~dnorm(mu[i],tau[i]) }
        # sampling of means
    tau[i] <- n[i] / pow(sigma,2)
}
m <- pow(4*sigma,2)/n[1]+pow(4*sigma,2)/(2*n[1])
for(k in 1:K) {    # bootstrap procedure
    t[k] <- (4*y[1,k]-(y[2,k]+y[3,k]+y[4,k]+y[5,k]))/pow(m,0.5)
    # Student's test
    spower[k] <- step(t[k]-clr)
}
r~dunif(1,K)
Student <- t[r]
Power <- mean(spower[])
nTotal <- sum(n[])
clr~dunif(cl[1], cl[2])
}
Data list(mu=c(35.6, 33.7, 30.2, 29.0, 25.9), sigma=3.75, cl=c(2,3),
        K=100)
inits <- function(){
    list(N=50, y=structure(.Data=rep(30,500), .Dim=c(5, 100)),
                    r=50, clr=2.5) }
```

It is helpful to exploit R code for inits, for in BUGS it is clumsy with 500 repetition of 30 in array y. I used Jeffreys priors for N, the sample size of each of group supplied with EZD1, EZD2, LZ1, LZ2. The total sample size is nTotal=sum(n_i)=6*N. The bootstrap is used to generate K subsamples of group means given N. The rational is not only to produce vector of power indicators (array spower[]), but also to make robust sampling of test statistic using r~dunif(1,K) drawing. The further robustness was supported by drawing test significance level roughly from 0.05-0.001 interval: clr~dunif(2, 3). Distributions of Power and Student's test values against the total sample size are given in Figure 1 (values of test above 5 are excluded for clarity).

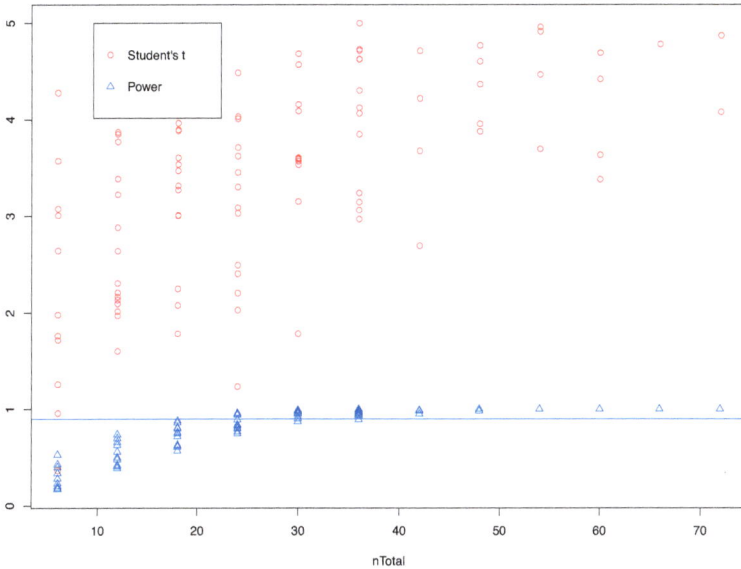

Fig. 1. Power and Student's test values against the total sample size in the ANOVA example

The proposed approach supplied us with the distributions of possible values rather than point estimates, extending terms of deductions. For example, in the SAS output, the sample size N Total is 30 with the corresponding power 0.947 given α=0.025.

We have Student's t and power distribution quantiles at nTotal=30 displayed in Table 4. These suggest that only 75% of possible measurements will indeed meet criteria.

Table 4. Student's t and power distribution quantiles in the One-Way ANOVA example

Nodes	Min.	1st Qu.	Median	Mean	3rd Qu.	Max.
Student's t	1.781	3.568	3.604	3.952	4.568	5.611
Power	0.870	0.900	0.950	0.935	0.970	0.990

The advantages of the proposed realization are obvious. Elaborations are numerous and easy to tackle. They may concern (i) mechanism of means generation (say, $y_i \sim \text{Student}(\mu_i, \tau_i, n_i)$), (ii) test (e.g. Fisher, Kruskal-Wallis, etc.), (iii) uncertainty about expected means with possible measurement error adjustment, (iv) sensitivity variations (consider options of `Student <- ranked(t[], K/4)` or `spower[k] <- step(t[k]-cl[1])`). Important extension is the opportunity to combine prior information with results of the pilot study, shaping the data based likelihood.

Extension 1. I demonstrate the elaboration on uncertainty about expected means. All one has to do is to replace

```
for(k in 1:K) {y[i,k]~dnorm(mu[i],tau[i])}
```

with

```
for(k in 1:K) {y[i,k]~dnorm(mutrue[i],tau[i])}
mu[i]~dnorm(mutrue[i],1)
```

So, now we assume that the true means may deviate from the expected one by 3 with 67% assurance that deviations are no more than 1. Distributions of Power and Student's test values against the total sample size are given in Figure 2 (values of test above 5 are excluded for clarity).

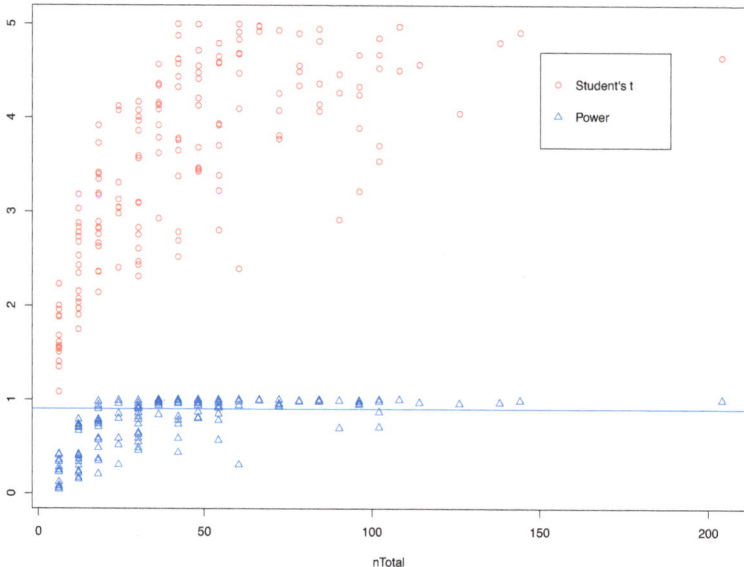

Fig. 2. Power and Student's test values against the total sample size in the ANOVA example with measurement error

I tabulated Student's t and power distributions quantiles across nTotal $\in$ {30, 36, 42} in Table 5

Table 5. Student's t and power distribution quantiles across nTotal $\in$ {30, 36, 42} in the One-Way ANOVA example with measurement error

Nodes	Min.	1st Qu.	Median	Mean	3rd Qu.	Max.
t (30)	1.662	3.059	3.618	3.829	4.743	5.961
Power (30)	0.560	0.770	0.970	0.886	1.000	1.000
t (36)	2.159	2.889	3.921	3.918	4.838	6.062
Power (36)	0.650	0.860	0.980	0.917	1.000	1.000
t (42)	2.369	4.713	4.875	5.105	5.630	7.174
Power (42)	0.900	0.990	1.000	0.983	1.000	1.000

We can clearly see the drop in power and precision given nTotal so the total sample size required stretched to 36 at least.

Extension 2. To test several contrasts together one should combine the tests into power calculus as demonstrated in the code below.

```
model {
    N~dcat(p[])
    for (i in 1:100) { p[i] <- 1/100 }
    for (i in 2:5) { n[i] <- N }
    n[1] <- 2*N
    for (i in 1:5) {
        for (k in 1:K) {y[i,k]~dnorm(mu[i],tau[i])}
        tau[i] <- n[i]/pow(sigma,2)
    }
    m1 <- pow(4*sigma,2)/n[1]+pow(4*sigma,2)/(2*n[1])
    m2 <- 2*pow(2*sigma,2)/(2*n[2])
    m3 <- 2*pow(sigma,2)/n[2]
    for (k in 1:K) {
        t1[k] <- (4*y[1,k]-(y[2,k]+y[3,k]+y[4,k]+y[5,k]))/pow(m1,0.5)
        t2[k] <- (y[2,k]+y[3,k]-y[4,k]-y[5,k])/pow(m2,0.5)
        t3[k] <- (y[2,k]-y[3,k])/pow(m3,0.5)
        t4[k] <- (y[4,k]-y[5,k])/pow(m3,0.5)
        spower[k] <- step(t1[k]-clr)*step(t2[k]-clr)*step(t3[k]-clr)*
                     step(t4[k]-clr)
    } ## combined power check
    r~dunif(1,K)
    Student1 <- t1[r]
    Student2 <- t2[r]
    Student3 <- t3[r]
    Student4 <- t4[r]
    Power <- mean(spower[])
```

```
    nTotal <- sum(n[])
    clr~dunif(cl[1], cl[2])
}
data list(cl=c(2,3), mu=c(35.6, 33.7, 30.2, 29.0, 25.9),
          sigma=3.75, K=100)
inits list(N=50, y=structure(.Data=rep(30,500),
           .Dim=c(5, 100)), r=50, clr=2.5)
```

Distributions of Power and Student's test values against the total sample size are given in Figure 3 (values of test above 15 are excluded for clarity).

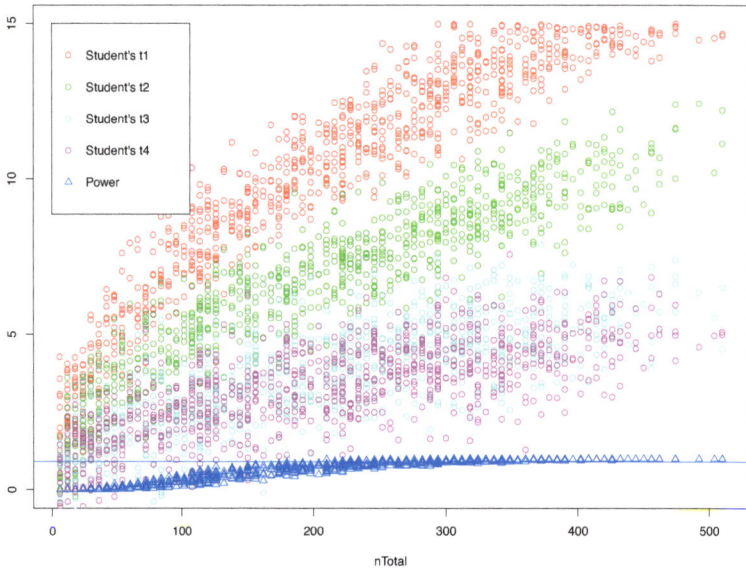

Fig. 3. Power and Student's test values against the total sample size in the ANOVA example, multiple contrasts

I extracted Student's t and power distribution quantiles across nTotal in the intervals [290, 310] and [390, 410] (latter displayed in Table 6).

From the results, the total sample size of 400 is sufficient to test all 4 contrasts simultaneously given the power above 0.9 and the alpha level 0.05/4=0.0125.

Elaboration on uncertainty about expected means. Again, I inserted additional code: $mu[i] \sim dnorm(mutrue[i],1)$, resulting in the results shown in Figure 4. The obvious pattern of the results is the requirement of larger sample sizes and the slower increase in power with the sample size.

Table 6. Student's t and power distribution quantiles across nTotal in the interval [390, 410] in the One-Way ANOVA example with multiple contrasts

Nodes	Min.	1st Qu.	Median	Mean	3rd Qu.	Max.
Student's t1	12.90	14.08	14.85	14.80	15.44	16.47
Student's t2	6.983	9.106	9.758	9.783	10.545	12.040
Student's t3	3.461	4.473	5.277	5.295	5.944	7.232
Student's t4	2.489	4.195	4.681	4.662	5.188	6.722
Power		0.930	0.970	0.990	0.980	0.990	1.000

For better illustration, I extracted Student's t and power distributions quantiles across nTotal in the intervals [400, 450] and [550, 600] (latter displayed in Table 7).

Table 7. Student's t and power distribution quantiles across nTotal in the interval [550, 600] in the One-Way ANOVA example, multiple contrasts with multiple contrasts

Nodes	Min.	1st Qu.	Median	Mean	3rd Qu.	Max.
Student's t1	10.69	15.17	17.68	17.80	20.07	26.82
Student's t2	4.700	9.563	11.740	11.466	13.535	17.550
Student's t3	0.838	5.108	7.396	7.097	9.138	14.280
Student's t4	-1.664	4.479	6.160	6.084	7.796	12.840
Power		0.000	0.803	1.000	0.823	1.000	1.000

Even taking the first quantile as appropriate, the lack of power against the required 0.9 level is obvious given sample size of 550-600.

Another generalization is the possibility to accommodate to multifactor ANOVA designs with more complex contrasts using the likewise BUGS code and multivariate ANOVA designs with the Hotelling's t-squared statistic based test.

2.2 Example 2. Comparing Two Survival Curves

This example is taken from SAS/STAT 9.1 User's Guide (2004), *Chapter 57. The POWER Procedure*. Example 57.6. Comparing Two Survival Curves (p. 3561) (SAS Institute Inc., 2004).

A clinical trial is used to compare survival rates for the proposed and standard cancer treatments. The planned data analysis is a log-rank test to compare nonparametrically the overall survival curves for the two treatments. The goal is to determine an appropriate sample size to achieve a power of 0.8 for a 2-sided test with $\alpha = 0.05$ using a balanced design. The survival curve for patients on the standard treatment is well known to be approximately exponential with a median survival time of five years. It is expected that the newly proposed treatment will yield a (nonexponential) survival curve with survival probabilities 0.95, 0.9, 0.75, 0.7, 0.6 in years 1 to 5.

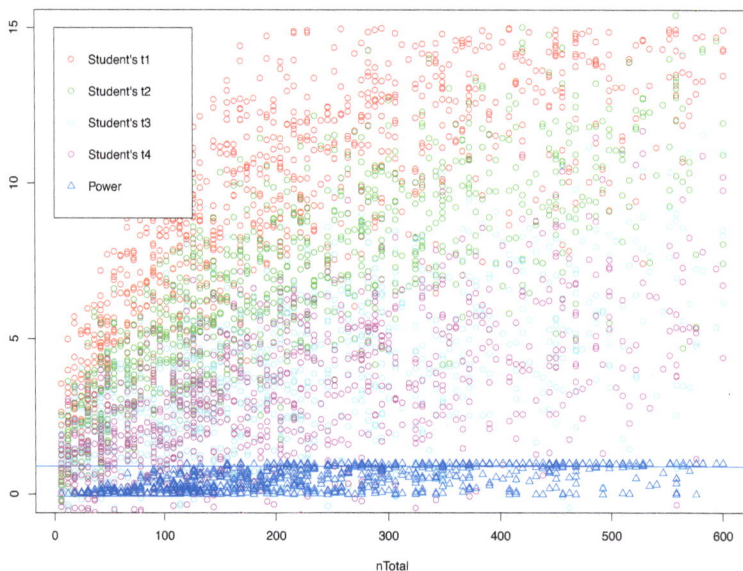

Fig. 4. Power and Student's test values against the total sample size in the ANOVA example, multiple contrasts with measurement error

Some loss to follow-up is expected, with roughly an exponential rate that would result in about 50% loss with the standard treatment within 10 years. The loss to follow-up for the proposed treatment is more difficult to predict, but 50% loss would be expected to occur sometime between years 5 and 20.

The following statements perform the analysis using SAS.

```
proc power;
twosamplesurvival test=logrank
curve("Standard") = 5 : 0.5
curve("Proposed") = (1 to 5 by 1):(0.95 0.9 0.75 0.7 0.6)
groupsurvival = "Standard" | "Proposed"
accrualtime = 2
followuptime = 3
groupmedlosstimes = 10 | 20 5
power = 0.8
npergroup = .;
run;
```

The output is given in Table 8.

The required sample size per group to achieve a power of 0.8 is 228 if the median loss time is 20 years for the proposed treatment. Only six more patients are required in each group if the median loss time is as short as five years.

Table 8. SAS output, Survival curves example

Index (Group)	Median Loss Time	Actual Power	N per Group
1	20	0.800	228
2	5	0.801	234

Proposed Method For the generation of means, I used the truncated normal survival probabilities sampling $S_i \sim \text{Normal}(\mu_i, \sigma^2/n_i)I[S_{i+1}, S_{i-1}]$ with $S_0 = 1$.

The test is the Z test using the standard normal distribution. The following notations will be used.

- $S_j(i)$ = input survivor function value corresponding to group j and time i;
- $h_j(t)$ = hazard rate for group j at time t;
- $\Psi_j(t)$ = loss hazard rate for group j at time t;
- m = median survival time;
- $N_j(i)$ = expected number at risk at time i in group j;
- T = accrual and post-accrual follow-up time;
- r_i = rank for the ith time point.

The distribution of the test statistic Z is approximated by $N(0,1)$:

$$Z = \frac{\sum_{i=1}^{T} D_i r_i \left[\frac{\phi_i \theta_i}{1 + \phi_i \theta_i} - \frac{\phi_i}{1 + \phi_i} \right]}{\sqrt{\sum_{i=1}^{T} D_i r_i^2 \frac{\phi_i}{(1 + \phi_i)^2}}}$$

where

$$\theta_i = \frac{h_2(t_i)}{h_1(t_i)}, \quad \phi_i = \frac{N_2(t_i)}{N_1(t_i)}$$

$$D_i = h_1(t_i) N_1(i) + h_2(t_i) N_2(i)$$

$$h_1(t) = -\log(0.5)/m$$
$$h_2(t_i) = \frac{S_2(t_i) - S_2(t_{i+1})}{S_2(t_i)}$$

$$N_j(i+1) = N_j(i) [1 - h_j(t_i) - \Psi_j(t_i)]$$

$$r_i = \begin{cases} 1, & log-rank \\ N_1(i) + N_2(i), & Gehan \\ \sqrt{N_1(i) + N_2(i)}, & Tarone-Ware \end{cases}$$

So with $\theta_i \approx 1$, i.e., $h_1(t) \approx h_2(t)$, one expects $Z \approx 0$. For the case under consideration, $h_1(t) > h_2(t)$, so that $Z < 0$.

The BUGS script for the proposed method is given below.

```
model {
    nTotal~dunif(100,500)
    TL50~dunif(5,20)
    for (k in 1:K) {
        S[1,k] <- 1
```

```
      Low[k] <- 0.5*S[T+L,k]
          for (i in 1:4) {
          S[i+1,k]~dnorm(mu[i], tau[i,k])I(S[i+2,k],S[i,k])
          INST[i,k] <- N2[i,k]/(mu[i]*(1-mu[i]))
          tau[i,k] <- max(50,INST[i,k])
          }
      S[T+L+1,k]~dnorm(mu[T+L],tau[T+L,k])I(Low[k], S[T+L,k])
      INST[T+L,k] <- N2[T+L,k]/(mu[T+L]*(1-mu[T+L]))
      tau[T+L,k] <- max(50,INST[T+L,k])
}
for(i in 1:(T+L) {
    h1[i] <- -log(0.5)/5
    loss1[i] <- -log(0.5)/10
    for(k in 1:K) {
        h2[i,k] <- step((S[i,k]-S[i+1,k])/S[i,k]-0.0001)*
                   (S[i,k]-S[i+1,k])/S[i,k]+equals(S[i,k]
                   -S[i+1,k],0)*h1[i]
        loss2[i,k] <- -log(0.5)/TL50
    }
}
N1[1] <- 0.5*nTotal
for (i in 2:T+L) {
   N1[i] <- N1[i-1]*(1-h1[i-1]-loss1[i-1])
   }
for (k in 1:K) {
    N2[1,k] <- 0.5*nTotal
    for (i in 2:T+L) {
        N2[i,k] <- step(1-h2[i-1,k]-loss2[i-1,k])*N2[i-1,k]*
                   (1-h2[i-1,k]-loss2[i-1,k])
    }
    for (i in 1:T+L) {
        theta[i,k] <- h2[i,k]/h1[i]
        phi[i,k] <- N2[i,k]/N1[i]
        D[i,k] <- (h1[i]*N1[i]+h2[i,k]*N2[i,k])
        M_N[i,k] <- phi[i,k]*theta[i,k]/(1+phi[i,k]*theta[i,k])
                    -phi[i,k]/(1+phi[i,k])
        M_D[i,k] <- phi[i,k]/pow(1+phi[i,k],2)
    }
Z[k] <- inprod(D[,k],M_N[,k])/pow(inprod(D[,k],M_D[,k]),0.5)
spower[k] <- step(clr-E[k])
}
r~dunif(1,K)
ZTest <- Z[r]
Power <- mean(spower[])
}
```

```
data list(clr=-1.8, T=2, L=3,
     mu = c(0.95, 0.9, 0.75, 0.7, 0.6), K=100)
inits list(nTotal=200, TL50=10,
     S=structure(.Data=rep(c(NA, 0.9,0.8,0.7,0.6,0.5), 100),
     .Dim=c(6, 100)))
```

Therefore, we can compare both the known distribution and empirically defined curves. There are numerous refinements possible to take care of feeding the sample with patients in accrual period given the empirical or known distribution, e.g., uniform distribution under constant inflow rate. I obviated it purposely to demonstrate the gist of the proposition. The great thing about the MCMC approach is its flexibility of rendering information insufficiencies. For example, vague understanding of the experimental cohort dwindling (50% loss would be expected to occur sometime between years 5 and 20) can be rendered by the uniform distribution under the constant cohort exhaustion rate coded TL50~dunif(5,20) in the script. Distributions of Power and Z test values against the total sample size are given in Figure 5.

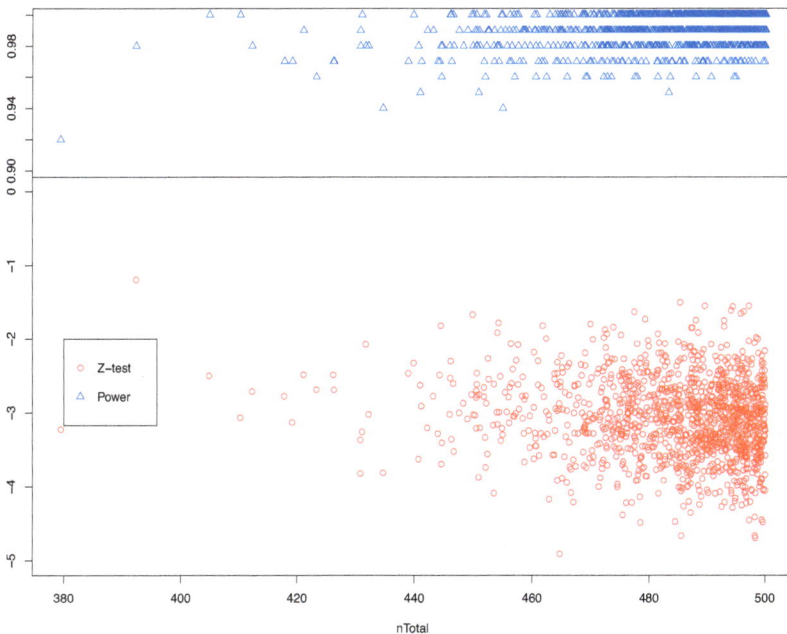

Fig. 5. Distributions of Power and Z test values against the total sample size in two survival curves example

The "5 numbers" excerpts of the output are tabulated in Table 9.

Table 9. Power and Z test values distribution quantiles across nTotal in the interval [440, 460], Survival Curves example

Nodes	Min.	1st Qu.	Median	Mean	3rd Qu.	Max.
Z (440-460)	-4.09	-3.20	-2.91	-2.89	-2.61	-1.67
Power (440-460)	0.94	0.98	0.98	0.98	0.99	1.00
Z (480-500)	-4.70	-3.47	-3.10	-3.11	-2.76	-1.50
Power (480-500)	0.95	0.99	0.99	0.99	1.00	1.00

Notice there is no equivalence in behavior of errors of the two types, one always more rigid to decrease over the other.

It seems natural to extend the planning to several cohorts design by applying the same tests contrasting survival curves with that of the basic (control) cohort. It is also possible to include information on covariates in hazard functions.

Extension. Application of Cochran-Mantel-Haenszel (CMH) test. The CMH test is more robust, so expectedly more demanding.

The CMH statistic has a chi-square distribution with $df = T$.

We fist define the following notations:

- a_i = number of events in experimental cohort in time i;
- $E(a_i)$ = expected number of events in experimental cohort in time i given $H0$;
- $V(a_i)$ = dispersion of a_i;
- D_i = total number of events in experimental and control cohort in time i;
- E_i = expected number at risk in experimental cohort at time i;
- $\bar{D}_i$ = number of survived in experimental group;
- $\bar{E}_i$ = expected number at risk in control cohort at time i; and
- n_i = total expected number at risk in control cohort at time i.

Then CMH statistic is:

$$CMH = \frac{\left(\sum_{i=1}^{T} a_i - \sum_{i=1}^{T} E(a_i)\right)^2}{\sum_{i=1}^{T} V(a_i)}$$

$$E(a_i) = \frac{D_i E_i}{n_i}; \quad V(a_i) = \frac{D_i \bar{D}_i E_i \bar{E}_i}{n_i^2 (n_i - 1)}.$$

The proposed method can be realized in the following BUGS script.

```
model {
    nTotal~dunif(200,900)
    TL50~dunif(5,20)
    for (k in 1:K) {
        S[1,k] <- 1
        Low[k] <- 0.5*S[T+L,k]
        for (i in 1:4) {
            S[i+1,k]~dnorm(mu[i], tau[i,k])I(S[i+2,k],S[i,k])
```

```
            INST[i,k] <- N2[i,k]/(mu[i]*(1-mu[i]))
            tau[i,k] <- max(50,INST[i,k])
        }
        S[T+L+1,k]~dnorm(mu[T+L],tau[T+L,k])I(Low[k], S[T+L,k])
        INST[T+L,k] <- N2[T+L,k]/(mu[T+L]*(1-mu[T+L]))
        tau[T+L,k] <- max(50,INST[T+L,k])
    }
    for (i in 1:(T+L)) {
        h1[i] <- -log(0.5)/5
        loss1[i] <- -log(0.5)/10
        for (k in 1:K) {
            h2[i,k] <- step((S[i,k]-S[i+1,k])/S[i,k]-0.0001)*
                (S[i,k]-S[i+1,k])/S[i,k]+equals(S[i,k]
                    -S[i+1,k],0)*h1[i]
            loss2[i,k] <- -log(0.5)/TL50
        }
    }
    N1[1] <- 0.5*nTotal
    for (i in 2:T+L) { N1[i] <- N1[i-1]*(1-h1[i-1]-loss1[i-1]) }
    for (k in 1:K) {
        N2[1,k] <- 0.5*nTotal
        for(i in 2:T+L) {
            N2[i,k] <- step(1-h2[i-1,k]-loss2[i-1,k])*N2[i-1,k]*
                    (1-h2[i-1,k]-loss2[i-1,k])
        }
        for (i in 1:T+L) {
            a[i,k] <- h2[i,k]*N2[i,k]
            D[i,k] <- (h1[i]*N1[i]+h2[i,k]*N2[i,k])
            Ea[i,k] <- D[i,k]*N2[i,k]/(N1[i]+N2[i,k])
            Va[i,k] <- D[i,k]*(N1[i]+N2[i,k]-D[i,k])*N1[i]*N2[i,k]/
                    (pow(N1[i]+N2[i,k],2)*(N1[i]+N2[i,k]-1))
        }
    chisq[k] <- pow(sum(a[,k])-sum(Ea[,k]),2)/sum(Va[,k])
    spower[k] <- step(chisq[k]-clr)
    }
r~dunif(1,K)
ChiTest <- chisq[r]
Power <- mean(spower[])
}
```

The output is displayed in Figure 6.

For illustration, I extracted the chi-square test and power distributions quantiles across nTotal in the intervals [740, 760] and [790, 810] (see Table 10).

One can, therefore, balance between robustness and sample size demand.

Table 10. Power and chi-square test values distribution quantiles across nTotal in the intervals [740, 760] and [790, 810], Survival Curves example

Nodes	Min.	1st Qu.	Median	Mean	3rd Qu.	Max.
		nTotal in interval 740-760				
χ^2 test	8.285	13.430	17.520	16.730	20.230	23.270
Power	0.870	0.900	0.930	0.925	0.950	0.970
		nTotal in interval 790-810				
χ^2 test	8.176	14.330	17.620	18.290	21.740	33.910
Power	0.900	0.930	0.950	0.947	0.960	0.990

2.3   Example 3. Log-Linear Models

This example is based on Congdon (2005, p. 139) just to demonstrate the application of the approach in generalized linear modeling.

The data y_{ijk} cross-classify the thromboembolism and control patients ($i= 1$ and 2, respectively) by two risk factors: oral contraceptive user ($j= 1$ for users, $j = 2$ for non-users) and smoking ($k =1$ for smokers, $k =2$ for non-smokers). The data are given in Table 11.

Table 11. Thromboembolism data

Patient type	Smoker		Non-smoker		Total
	contraceptive user		contraceptive user		
	Yes	No	Yes	No	
Thromboembolism	14	7	12	25	58
Control	2	22	8	84	116

While a design matrix approach to model specification as in Bishop et al. (1975) is often used, an economical model notation following Worcester (1971) is used here, with the saturated model specified as

$$y_{ijk} \sim Po\left(\mu_{ijk}\right)$$
$$\log\left(\mu_{ijk}\right) = \beta_0 + \delta_i\beta_1 + \delta_j\beta_2 + \delta_k\beta_3 + \delta_{ij}\beta_{12} + \delta_{ik}\beta_{13} + \delta_{jk}\beta_{23} + \delta_{ijk}\beta_{123}$$

where δ is one only when all the subscripts are one, and is zero otherwise. Thus $\delta_1 = \delta_{11} = \delta_{111} = 1$ but for all other subscript combinations $\delta = 0$. Hence, the need to specify corner constraints is eliminated.

For the data, a close fit was obtained with a model omitting β_{13} and β_{23}. For clarity, I purposely omitted β_{123}, too. The hypothesis under concern was the presence of significant positive association between thromboembolism and oral contraceptive usage, that is $\beta_{12} > 0$. In most cases, the second order effects that are of concern. Therefore, I focused on the following model specification

$$\log\left(\mu_{ijk}\right) = \beta_0 + \delta_i\beta_1 + \delta_j\beta_2 + \delta_k\beta_3 + \delta_{ij}\beta_{12}.$$

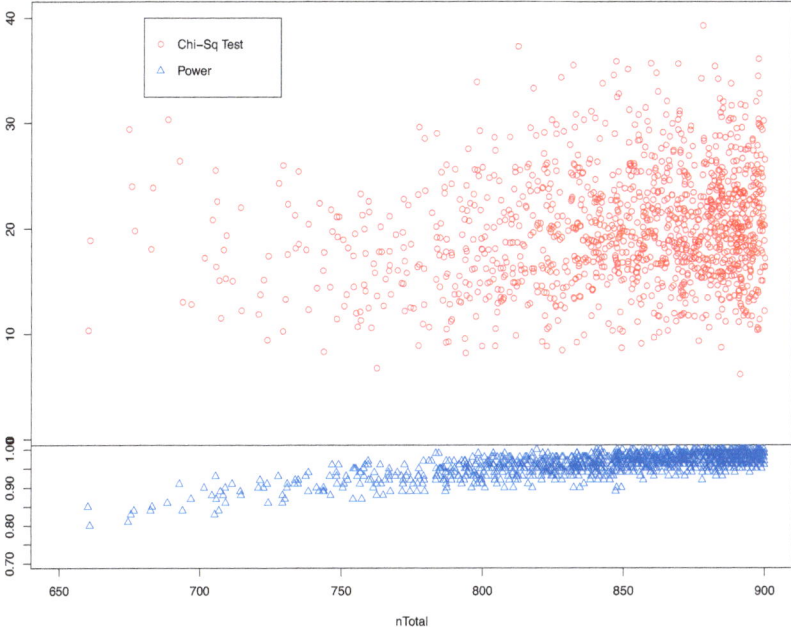

Fig. 6. Distributions of Power and Cochran-Mantel-Haenszel test values against the total sample size in the example of two survival curves

In the script, the four regression coefficients constitute the vector beta so that beta[5] is of concern. The only predicament is to obtain the variance of β_{12}.

Having the MCMC chains, it seems to be easy but unfortunately chains (and so posterior distributions of parameters) in BUGS are not accessible after the analysis is carried out. We can, of course, conduct a set of analyses at different sample sizes to have the appropriate variances or use R packages such as asypow. I preferred to use the Fisher information matrix that one can use efficiently for the purpose in GLM. The Fisher information matrix for log-linear models is conveniently delivered by R packages such as Palmgren (1981). Here, I set up the model in terms of the original setup (Table 12 corresponds to Table 1 in Palmgren (1981)) as shown below.

Table 12. Structure of means under the Poisson model

Outcome \ exposure	Means under Poisson	
	contraceptive +	contraceptive -
Thromboembolism	$\mu_{11} = \exp(\lambda 1 + \beta + \rho)$	$\mu_{12} = \exp(\lambda 2 + \beta)$
Control	$\mu_{21} = \exp(\lambda 1)$	$\mu_{22} = \exp(\lambda 2)$

$\rho -$ beta[5] (parameter of concern, association)

$\beta = \text{beta}[2]$ (presence of thromboembolism)
$\lambda 1 = \text{beta}[1] + \text{beta}[3] + \text{beta}[4]$
$\lambda 2 = \text{beta}[1] + \text{beta}[4]$
beta[1] - constant term
beta[3] - contraceptive user
beta[4] - smoker

It has been shown that the diagonal element of the Fisher information matrix corresponds to ρ tantamount to (n_1 is the marginal frequency of the contraceptive users):

$$\frac{-\mu_{11}\mu_{12}}{n_1}.$$

Therefore, the precision of beta[5] (tau in the script) equals $8\mu_{11}\mu_{12}/n_1$.

The basic design is chosen so that the minimal cell frequency should be equal to one. I checked the dispersion values derived analytically with posterior BUGS estimates at different basic design multipliers. With multiplier=2 (original data), the estimates are 0.245 and 0.204, with multiplier=10 the estimates are 0.05 and 0.04, with multiplier=40 they are 0.013 and 0.011, and with multiplier=200 they are 0.0026 and 0.0026. Indeed, estimates of the precision based on the Fisher matrix are quite reliable. The BUGS script for the proposed method is given below (all priors on beta parameters preserved from Congdon (2005)).

```
model {
    for (mult in 1:20) {
        for (i in 1:8) {
            y[i,mult] <- yy[i]*mult/2
            y[i,mult]~dpois(mu[i,mult])
            log(mu[i,mult]) <- beta[1,mult]+equals(a[i],1)*
                                beta[2,mult]+equals(b[i],1)*
                                beta[3,mult]+equals(c[i],1)*
                                beta[4,mult]+equals(a[i],1)*
                                equals(b[i],1)*beta[5,mult]
        }
        for (j in 1:5) { beta[j,mult] ~ dnorm(M[j],P[j]) }
        log(mu11[mult]) <- beta[1,mult]+beta[2,mult]+beta[3,mult]+
                            beta[4,mult]+beta[5,mult]
        log(mu21[mult]) <- beta[1,mult]+beta[3,mult]+beta[4,mult]
        tau[mult] <- 8*mu11[mult]*mu21[mult]/(y[1,mult]+y[3,mult]+
                        y[5,mult]+y[7,mult])
        for (k in 1:K) {
            bB[k,mult]~dnorm(beta[5,mult], tau[mult])
            t[k,mult] <- bB[k,mult]*sqrt(tau[mult])
            spower[k,mult] <- step(t[k,mult]-2)
        }
        Student[mult] <- t[r,mult]
        Power[mult] <- mean(spower[,mult])
```

```
        nTotal[mult] <- sum(y[,mult])
    }
r~dunif(1,K)
}
data list(yy=c(14,7,12,25,2,22,8,84), a=c(1,1,1,1,2,2,2,2),
         c=c(1,1,2,2,1,1,2,2), b=c(1,2,1,2,1,2,1,2),
         M=c(0,2,2,2,2,2,2,2),
         P=c(0.001,0.1,0.1,0.1,0.1,0.1,0.1,0.1),
         K=100)
inits list(beta=structure(.Data=rep(c(-1,1,-1,1,-1), each=20),
                          .Dim=c(5, 20)),
           bB= structure(.Data=rep(0, 100*20),
                         .Dim=c(100, 20)), r=50)
```

A part of the WinBUGS output is given in Table 13.

Table 13. Part of the WinBUGS output for power calculation

node	mean	sd	MC err.	2.5%	median	97.5%
Power[1]	0.728	0.230	0.0073	0.22	0.77	1
Power[2]	0.935	0.106	0.0031	0.60	0.98	1
Power[3]	0.990	0.028	0.0009	0.91	1	1
Power[4]	0.998	0.012	0.0004	0.98	1	1
Power[5]	1.000	0.003	0.0001	0.99	1	1
Student[1]	2.913	1.479	0.0484	0.31	2.82	5.99
Student[2]	4.172	1.447	0.0401	1.39	4.12	7.08
Student[3]	5.249	1.473	0.0382	2.46	5.20	8.29
Student[4]	5.930	1.474	0.0494	3.18	5.89	8.79
Student[5]	6.579	1.420	0.0439	3.95	6.56	9.40

As it turned out, the basic design of the sample size of 87 gives a median power of 77%. To assure the median power above 80%, the original sample size of 174 is enough (Power[2]). To secure power above 90%, a sample size of 261 is needed (Power[3]), which is 1.5-fold of the sample size of original study.

3   Discussion

The median of the posterior distribution of beta[5] based on the data in Table 11 is 1.595 with 95% CI [0.674, 2.562]. Is it not strong enough evidence of association? Why should we collect 1.5-fold as many data? The answer of course is related to the specifics of the sample. The basic design itself is a sample, not status quo that represents true frequencies ratios in population. Therefore, we have to assure that the sample data bring in enough information to obscure sample specifics. Of course, the more complex the design is and the more sample variation has to be outbalanced by signal, the larger sample size is required. The original data in Table 11 is one of the random snapshots of reality and we have to put as much credit as possible to it. Not all snapshots of size 174 guarantee a 95% CI includes zero. Two point five percent of them elicit power less than 60% while the sample size of 260 affords enough power to assure the significance of the association in almost all samples. The same logic is behind any application of power analysis. The other lay belief is that with the increase of sample size any association is doomed to be significant. For sure, it is not, and the strength of power analysis is to determine the optimal sample size of hypothesis testing. The sample size assures that there is no prospect of decisive augmentation of power and significance following the increase in sample size.

We also want to bring the attention to the data generation mechanism (DGM). Each hypothesis defines some environment and research design is elaborated to reproduce the essentials of this environment. The likelihood and in part priors include elements of the design. If a design poorly reproduces the environment, the DGM also reproduces poorly. Furthermore, each design mottles with predispositions to particular biases. So what we really capture is a biased DGM. I call it measurements generation mechanism (MGM). The strength of the recommended approach is the possibility to improve MGM toward DGM by using different well-known techniques that restore correct signal. I managed to introduce measurement error in the examples, but even more challenging issues are the corrections for other biases, especially for pervasive selection biases using numerous available techniques (Woolridge, 2002). It is important that the approach is capable to accommodate to these corrections in power analysis.

References

Bishop, Y. M., Fienberg, S. E., & Holland, P. W. (1975). *Discrete multivariate analysis: Theory and practice.* MIT Press: Cambridge, MA.

Congdon, P. (2005). *Bayesian models for categorical data.* John Wiley & Sons.

Gelman., A. (2015). The R project for statistical computing [Computer software manual]. Retrieved from https://cran.r-project.org/web/packages/R2WinBUGS/index.html

Lunn, D., Jackson, C., Best, N., Spiegelhalter, D., & Thomas, A. (2012). *The BUGS book: A practical introduction to bayesian analysis.* Chapman and Hall/CRC.

Palmgren, J. (1981). The fisher information matrix for log linear models arguing conditionally on observed explanatory variable. *Biometrika, 68*(2), 563–566. doi: https://doi.org/10.1093.

SAS Institute Inc. (2004). *Sas/stat 9.1 user's guide, version 9.2.* SAS Publishing Cary, NC, USA.

Woolridge, J. M. (2002). Sample selection, attrition, and stratified sampling. In *Econometric analysis of cross section and panel data* (chap. 17). Cambridge, London: The MIT Press.

Worcester, J.   (1971).   The relative odds in the 2-3 contingency table.   *American journal of epidemiology*, *93*(3), 145–149.   doi: https://doi.org/10.1093/oxfordjournals.aje.a121240.

An Application of Aspect-Based Sentiment Analysis on Teaching Evaluation

Wen Qu and Zhiyong Zhang

University of Notre Dame, Notre Dame IN 46556, USA
wqu@nd.edu; zzhang4@nd.edu

Abstract. With the rapid development of new techniques, text mining has become explosively popular in the past two decades. Various techniques and methods have been developed to manage and analyze text data to exploit the information underlying the text. Among them, the aspect-based sentiment analysis (ABSA), which is a research field that studies people's opinion, sentiment toward attributions or aspects of individual entities, has attracted researchers in both industry and academia. ABSA first extracts the relevant aspects of a specific entity and then determines the sentiment for each aspect. To our knowledge, there is no ready-to-use R packages or functions for ABSA. In this study, a brief review of ABSA is conducted and applied to a teaching evaluation study. It is also illustrated how to conduct ABSA using R.

Keywords: Aspect-based sentiment analysis · Teaching evaluation · Text data.

DOI: 10.35566/isdsa2019c6

1 Introduction

In daily life, when we need to make a decision, we often consider others' opinions, such as purchasing a product, searching for a restaurant, and choosing a doctor. With the rapid growth of the Internet, almost every aspect of life has been changed dramatically. How to make decisions is also changing. Instead of asking family members and friends for advice, people tend to use social media for help nowadays. For example, when someone wants to buy a new computer, he or she may search for comments on the websites of online retailers. One can obtain valuable information from others who have a similar experience by simply glancing over a few number of comments.

However, when there are many comments and reviews, it can be difficult for people to summarize information quickly. How to extract useful information from the texts is difficult but critical, especially in the digital age. Text mining, therefore, becomes very popular in the recent two decades with the growth of social media (Liu, 2015). Sentiment analysis is one of the various techniques and methods for managing and analyzing text data to exploit the underlining information.

Text mining is a field that analyzes people's opinions and sentiments towards certain entities, and their aspects (or attributes) expressed in the text (Liu & Zhang, 2012). It got popular after 2000 and was originated from computer science, but its applications have spread to business, management, sociology, political science, and literature. Moreover, its research has been mainly carried out at three levels: document, sentence, and aspect. In this study, we limit the scope to the last one. Comparing to both document and sentence level analyses, the aspect-based sentiment analysis (ABSA) can return the sentiment of aspects or the attributes of an entity instead of the overall polarity of the entity. In other words, we get to know the specific target of opinion with ABSA.

For an ABSA, with a particular entity, there are two main tasks: first, to extract the aspects from the text that needs to be evaluated, and second, to classify the sentiment for each aspect. In the literature, the aspect extraction task involves expression extraction and grouping. In Liu's (Liu, 2015) book, he gave a comprehensive review of the well-known methods and their applications. Due to the page limit, we only review the related methods that have been used in our study. The most straightforward approach to extract aspect expression is through the frequency of nouns or noun phrases (Blair-Goldensohn et al., 2008; Hu & Liu, 2004). This method is simple but effective because the nouns are often used to describe the aspects, and the vocabulary people use tends to be similar when they talk about different aspects under the same entity. By applying some rules, researchers can keep specific frequent nouns as aspect expressions. With the expressions, unsupervised methods (i.e., using dictionaries to find synonymous expressions or expert labeling) are applicable for grouping the expressions when the entity requires less specific knowledge (e.g., general product review and teaching evaluation). The second task, sentiment classification, can be accomplished by supervised learning methods (Jiang, Yu, Zhou, Liu, & Zhao, 2011) or unsupervised lexicon-based methods (Taboada, Brooke, Tofiloski, Voll, & Stede, 2011).

The ABSA is a domain-sensitive method; that is, a well-developed method for some domains may not be suitable for a different domain. To our knowledge, there is no method available in the literature with regards to teaching evaluation. Depend on the type of audience, the evaluation of teaching might have different purposes (Ory, 2000), including gathering feedback for teaching improvement, collecting data for personnel decision-making, or providing options for course selection. In the digital age, the form of teaching evaluation switched from the paper to the online version with both quantitative data and qualitative comments. This change makes it possible to conduct a more comprehensive mixed analysis like the longitudinal studies or text mining with hundreds or thousands of evaluation records. Therefore, in the rest of the paper, we apply ABSA to a teaching evaluation study and illustrate how to conduct the analysis using an R function we developed.

2   Application

In this section, we show how to apply the ABSA to the text comments on teaching evaluation. We first describe the data used in this study. Then, we propose a two-stage procedure to conduct ABSA. After that, we illustrate the procedure through the real data.

2.1   Data

The data were crawled from the *ratemyprofessor.com* website by conforming to their crawling rules. The crawled dataset contains 1954 records of teaching evaluation from students for 50 college professors in the United States. Each record is an evaluation from one student on one professor. The sample size is considerably larger than typical ABSA studies.

The dataset includes five variables. The first variable is the identification number of a record. The second variable is the unique id of a professor. The third variable is the numerical rating of a professor with 1 to 5 indicating worst to best to the answer of a question — "How would you rate this professor as an instructor?" The fourth variable is the date of a comment. The time of the records ranges from the year 1999 to 2018. The last variable is the response to an open-ended question about overall evaluation of a professor. Every student who filled out the evaluation form was asked to write a short comment here.

The histogram of the number of evaluations received by each professors is given in Figure 1. The number ranges from 8 to 98, and on average, each professor received 39 evaluations (the median is 33).

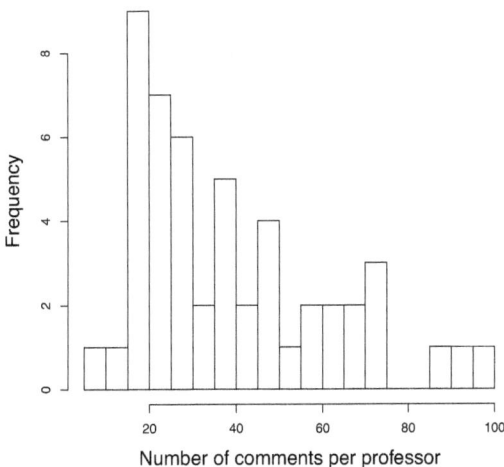

Fig. 1. Histogram of number of evaluations received by each professor

Two histograms of the rating scores are displayed in Figure 2. The left histogram is for the 1954 comments, and the right histogram is for the averaged rating score of each professor. With the grouping variable (here is the 'professor id'), the distribution of the rating scores show a bimodal pattern to a negative skewed pattern.

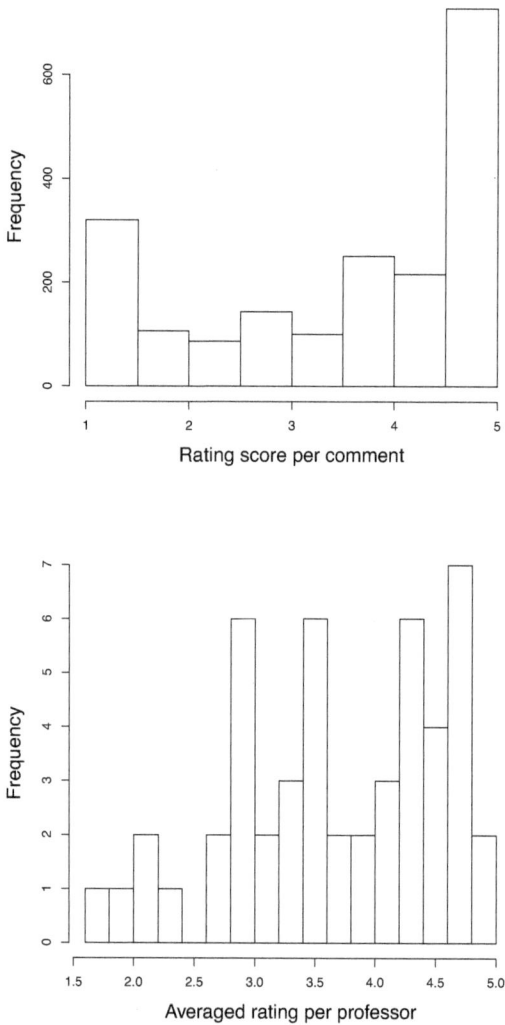

Fig. 2. Histograms of rating scores

The comments are in the free text format with a maximum of 350 characters for each. An example comment is given below.

> *Great teacher, really know his stuff. Also use a TON of example to try to explain everything well. Will give example of crime and let the class talk it out. You are free to ask question and he will answer very well. Also put his note on blackboard/hand them out in class, great to study with. I would suggest him.*

This comment clearly expressed a general positive sentiment towards the professor under evaluation. Our goal is to understand the sentiment of all 1,954 comments in the dataset.

2.2 Procedure

The procedure for evaluating teaching based on the text comments using ABSA involves two sequential stages. First, we discuss a combined method with word frequency, lexicon, and human labeling to extract aspects. Second, we develop an unsupervised scoring method for aspects' sentiment. The details are provided as follows.

Aspect Extraction and Grouping. The first task of ABSA is always to extract aspect expressions and then group them into aspect categories. We illustrate the procedure using the teaching evaluation data. We began by selecting all the nouns from all the comments, which led to a total of $n = 1511$ words. We then applied the following four constraints to obtain a shorter but more measurable and meaningful list:

(1) removing the stop words which are commonly used English words not having important meanings;
(2) removing nonsense nouns which are not meaningful in the context of the specific entity (in this example, the entity is teaching evaluation), such as 'bit', 'ton';
(3) adding topic specific words which are not in the previous word list, such as 'his','her';
(4) setting the frequency threshold to remove not frequently used words ($freq > 9$).

The remaining words were the candidates of the aspect expressions ($n = 189$) such as 'professor', 'homework', and 'quiz'.

Based on our knowledge on teaching evaluation, we decided to have five aspect categories: (A1) class in general, (A2) teacher in general, (A3) exam, (A4) grade, (A5) workload and class activity. The *class in general* aspect includes the expressions indicating class environment, course structure, materials, and the other general perspectives of a class. The *teacher in general* aspect describes the personality, teaching style, and research ability of a professor. The *exam, grade,* and *workload and class activity* aspects measure the other teaching evaluation related facets.

After having the expressions candidates and deciding the categories, we first labeled some expressions candidates as seed set and got their synonymous from the unlabeled set. We then manually labeled the rest based on the general knowledge about teaching evaluation. In total, we extracted 110 aspect expressions under five categories. Table 1 shows the aspect categories with their corresponding expressions.

Table 1. Aspect categories and expressions

Aspect Category	Size	Aspect Expressions
A1 (Class in general)	N1=52	class lecture mathematics hour book material text note stuff speech subject topic guide textbook chapter history powerpoint office chemistry econ lab art concept forum information syllabus music email video skill handout participate business instruction term expectation finance literature science approach classroom explain requirement yoga calculus line require movie schedule animal detail english
A2 (Teacher in general)	N2=30	professor teacher guy teaching person prof dr teach instructor experience style humor accent knowledge research writer prof. he she his her him himself herself he's she's
A3 (Exam)	N3=9	exam test essay final midterm paper quizzes quiz quizze
A4 (Grade)	N4=10	grade credit attendance grading attention level pass score gpa bonus
A5 (Workload & class activity)	N5=9	homework assignment time project reading writing practice discussion presentation

Aspect Scoring. In the current literature and applications, the majority of researchers focus on the orientation of the sentiment. To facilitate future analysis with other numerical variables (e.g., rating scores), we focus on the intensity of the sentiment—getting the sentiment scores for each aspect.

We used a four-step algorithm to score the aspects. First, for each comment, we scored the sentiment expression using the AFINN dictionary (a word list with discrete ratings of 2477 sentiment expressions). Concurrently, we added the aspect category label to each aspect expression (e.g., an aspect expression 'teacher' was labeled as its category 'A2'). Second, we applied the negation shifter (such as 'not', 'no', 'neither', 'nor') for sentiment ratings to change the direction of the negation sentiment expressions. Third, with aspect expressions and scored sentiment expressions, we applied the syntactic dependency rules to link them and assigned the score of a sentiment expression to its corresponding aspect expression. The syntactic dependency rules define the grammar relation between two words in a sentence with one word being the root and the other being the dependent. According to the Stanford Dependencies manual (de Marneffe & Manning, 2008) and our data, we selected 11 rules to quantify the dependency relationship between the aspect expressions and the sentiment expressions. Fourth, we aggregated the sentiment scores per aspect of each comment, and then got the averaged scores for each professor.

Example We use a simple example with only one sentence to demonstrate the procedure of the aspect scoring algorithm. The sentence is *"I love the prof, but the exam is not easy and long."* In the first step, we identified the aspect expressions with their categories (*prof*→A2, *exam*→A3) and scored the sentiment expressions (**love**→3, **easy**→1). Second, with the negation shifter, we changed the direction of the negative sentiments (**not easy**→-1). Third, we applied the syntactic dependency rules to link the sentiment expression with its aspect expression (*prof* $\overset{\text{obj}}{\Longrightarrow}$ **love**, *exam* $\overset{\text{nsubj}}{\Longrightarrow}$ **not easy**). Specifically, the *obj* rule indicates that the aspect expression *prof* is the direct object of sentiment expression **love**. The *nsubj* rule shows the aspect expression *exam* is the subject of the sentiment expression **easy** which has been changed the direction by the negation shifter ('not') in the previous step. Figure 3 shows the syntactic dependency relationship of the sentence.

In the last step, we aggregated the scores for each aspect category. In summary, in this example, only two aspects were involved in the sentence, and scores for them were 3 (A2: teacher aspect) and −1 (A3: exam aspect).

2.3   Results

In the current study, we analyzed all the comments to get the averaged aspect scores for each professor. Table 2 shows the scores of the 5 aspects, number of comments, and the averaged rating scores for the 50 professors. A positive/negative aspect score means the professor received an overall positive/negative opinion in that aspect among all comments he or she received. Because all the aspect scores were averaged, score 0 means the positive and negative opinions were canceled out. NA means the professor got no comment containing any expressions of that aspect. It usually happened when there were not enough comments like the professor with ID 3, who only had eight comments. If one aspect has a large amount of NA values, the researchers may change that aspect category, like adding more aspect expressions, combining with another category, or deleting the current one.

Different combinations of the aspect sentiments indicate distinctive teaching types. For example, professor 12 had 36 comments, and students positively evaluated his/her class structure and teaching style and course workload, but had a negative evaluation about the exam. Professor 40 had a similar amount of comments, but the evaluation was opposite. He/she received overall negative evaluations on the class, exam, grading, and workload aspects, but a positive evaluation on the teaching aspect.

Table 2: Aspect scores of 50 professors (N: number of comments, A1: Class, A2: Teacher, A3: Exam, A4: Grading, A5: Workload)

ID	N	Rating	Class	Teacher	Exam	Grading	Workload
1	28	4.25	0.29	1.67	-0.12	-0.5	-0.33
2	50	3.78	-0.19	1.43	-0.17	-0.08	0.14
3	8	3.69	-0.29	0.4	NA	0	NA
4	47	4.03	0.65	2.44	-0.33	0.75	0.28
5	72	2.94	-0.39	0.93	0.23	0.13	-0.11
6	98	3.83	0.19	1.23	0.15	-0.07	0.2
7	18	4.22	-0.56	1.69	1.25	1.5	0
8	17	3.41	-0.2	1.41	0.25	0.6	0.17
9	73	4.36	0.78	2.06	0	0.22	0.48
10	60	4.42	0.1	2.78	-0.44	0.14	-0.11
11	17	2.24	0.38	-0.25	1	-0.6	0
12	36	4.65	2.07	2.73	-0.91	0	0.33
13	66	3.47	-0.1	1.43	-0.02	-0.06	-0.06
14	21	2.1	-0.95	-0.28	-0.11	0.12	-0.5
15	27	4.28	1.64	2.04	0	-0.29	0.25
16	22	4.82	0.45	4.27	0.33	0	0
17	18	3.03	0.14	0.47	0	0	-0.33
18	18	4.5	0.18	2.31	-0.12	0	0
19	35	3.51	-0.32	1.71	0.19	0	0.58
20	74	2.14	-0.12	-0.22	0.17	0.11	-0.04
21	20	3.38	0.71	1.56	-0.54	-0.57	0.43
22	44	4.68	0.49	2.65	0.61	0.6	-0.5
23	47	4.49	2.02	2.15	-0.08	0.25	0.36
24	18	3.36	-0.29	0.83	-0.71	0.6	0
25	60	3.84	0.42	1.96	-0.19	-0.24	0.04
26	53	2.94	-0.02	0.64	0.14	0.38	0.17
27	28	3.5	-0.12	0.89	0.23	0.08	0.08
28	36	3.47	0.52	0.5	0.06	0.19	0.55
29	18	4.44	0.53	1.76	0	0.33	-0.5
30	61	4.13	0.41	1.3	0.18	-0.56	-0.27
31	38	4.36	1.57	2.84	-0.12	-0.14	0
32	39	3.32	0.05	1.41	-0.67	-0.13	0.07
33	22	2.86	0.06	-0.11	0	0	-0.5

Table continues on next page

Table 2: Aspect scores of 50 professors (N: number of comments, A1: Class, A2: Teacher, A3: Exam, A4: Grading, A5: Workload)

ID	N	Rating	Class	Teacher	Exam	Grading	Workload
34	17	2.71	-0.88	2.12	-0.08	0.14	0.4
35	12	2.92	0.5	-0.09	0	0	0
36	94	4.08	0.6	2.31	0.16	0.35	0.17
37	31	4.85	0.7	4.97	0	-0.25	0
38	28	4.7	1.88	3.37	0.29	-0.5	-0.5
39	49	4.66	0.83	1.77	0.05	0	0.16
40	37	2.62	-0.03	0.14	-1	-0.29	-0.44
41	86	1.71	-0.41	-0.52	-0.05	-0.47	-0.11
42	67	3.6	0.58	2.26	0.15	0.17	-0.06
43	26	4.73	0.33	3.12	0.07	0.73	0.15
44	21	3	0.24	0.33	0.08	-0.33	0.5
45	21	1.81	-0.06	-1.21	-0.14	-0.75	-0.12
46	22	3.11	0.19	0.59	0	-0.75	-0.43
47	41	3	0.3	1.06	0	0.1	0.1
48	65	4.77	1.43	3	0.14	0.07	0.08
49	26	4.71	0.52	2.96	0.5	0.33	0.25
50	22	4.36	0	2.86	0	0.3	0.17

The histograms and the correlations of the five aspects are shown in Figure 4. The class in general (A1) and teacher in general (A2) aspects are moderately positively correlated ($r = 0.49, p < 0.001$). Since the first and the second aspects contain most of the aspect expressions ($n_1 = 52, n_2 = 30$) and are more related to the teaching evaluation covered in the written comments, most of the students have similar opinions on these two aspects. For example, if a student thinks his or her professor is good at teaching (A2), very likely, he or she would also have a positive opinion of the class (A1). This is because a teacher's behavior usually profoundly affects the quality of the class in general.

The other aspects have weakly or none correlations, indicating that students rarely comment on those aspects together. Another explanation could be because only a few expressions were involved in the last three aspects. Thus, those aspects might be hard to be found in the comments. Future studies therefore need to consider adding more expressions in those aspects or redesigning the aspect categories.

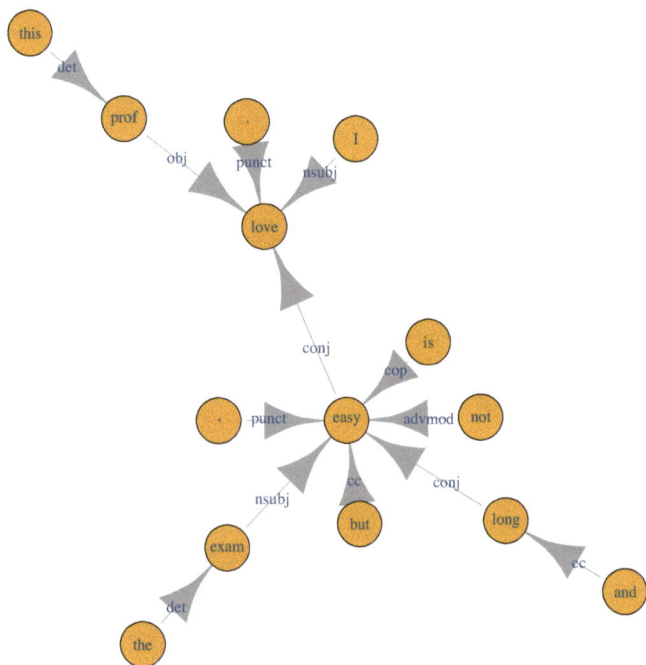

Fig. 3. The syntactic dependency relationship of one sentence (*I love the prof, but the exam is not easy and long.*)

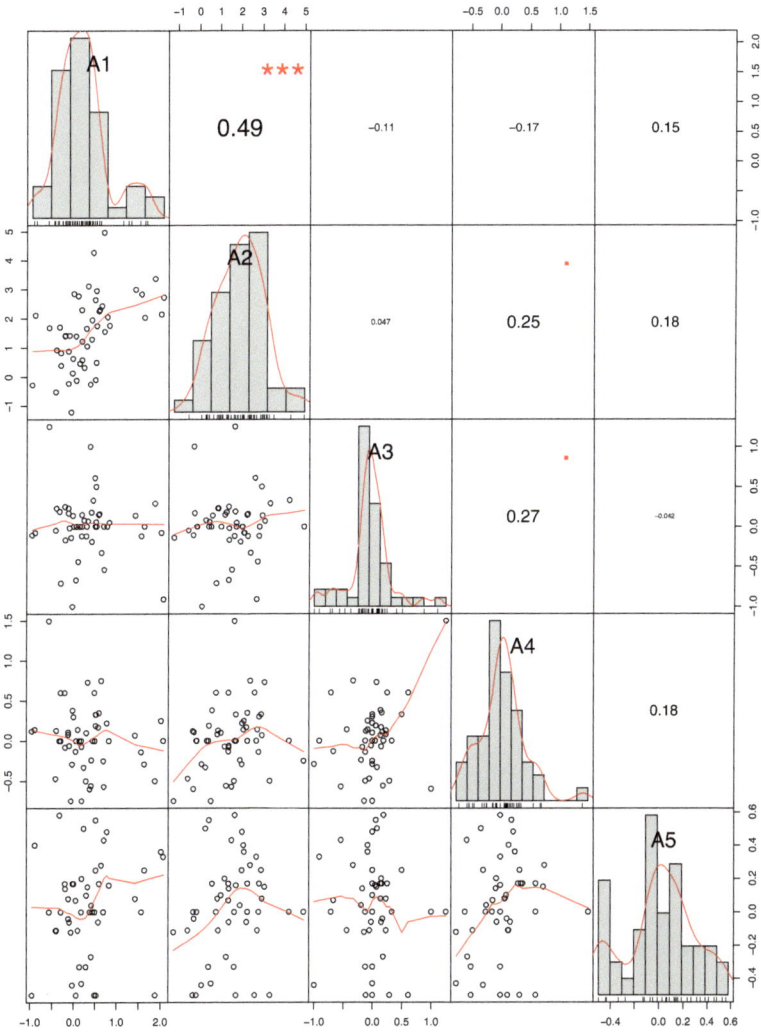

Fig. 4. Histograms and correlations of five aspects (* $p \leq 0.05$, ** $p \leq 0.01$, ***$p \leq 0.001$)

3   R Function

We also developed an R function to implement the aspect scoring algorithm, which is provided in the appendix. There are three arguments in the function:

- *comments*: A list of comments.
- *aspect.list*: A list of customized aspect categories with expressions.
- *cinf*: The additional information of the comments. If there is additional information (e.g., professor id, time), users can put them in a data frame and assign it to the cinf argument. The default value is NULL.

The output depends on whether there is additional information. If there is no additional information, the output would be the comments with the sentiment scores of each aspect category. An example of two comments with R input and output is given below.

```
> comment=c("Class is not easy but don't get discourage after the
midterm. The final is not bad. He is a good prof, but the lecture
is useless. Focus on the materials.", "I love this prof, but the
exam is not easy and long." )
> example=aspect_scoring(comment,aspect_hl)
Joining, by = "word"
Joining, by = "word"
> example
  comment_id A.1 A.2 A.3
1          1  -3   3   3
2          2  NA   3  -1
```

From the result, we found that the first comment included three teaching evaluation aspects — class (A1), teacher (A2), and exam (A3). The sentiment scores were consistent with the comment, in which the student explained the class was not easy, but the exam and teacher were good. The second comment was from the example in section 2.2. Because the second one only covered two aspects, the aspect not present in the comment had the score as NA.

On the other hand, if there is additional grouping information, like our data (grouping variable: professor ID), the output would return a table of professors with averaged sentiment scores of each aspect category averaging over the corresponding comments (like in Table 2). Users can easily adjust this option by modifying the R function to fulfill their need to obtain the averaged aspect scores based on people, time, institute, and other grouping variables.

4   Conclusion

In this study, we developed an approach for teaching evaluation using ABSA. We generated an aspect list for this domain and provided an R function for aspect sentiment scoring. By showing the procedure step by step and providing the R function, we hope to shine a light on the ABSA in the teaching evaluation field.

With the aspect sentiment scores, other data analysis can be conducted, e.g., a mixed analysis of both text data and quantitative data.

In the future, instead of using the existing sentiment dictionary such as AFINN, we will develop a domain-specific lexicon for better sentiment expression identification. Furthermore, we plan to gather information from both students and teachers about the teaching evaluation for a more precise aspect elicitation.

References

Blair-Goldensohn, S., Hannan, K., McDonald, R. T., Neylon, T., Reis, G. A., & Reynar, J. (2008). Building a sentiment summarizer for local service reviews..

de Marneffe, M.-C., & Manning, C. D. (2008). Stanford typed dependencies manual [Computer software manual]. Retrieved from https://nlp.stanford.edu/software/dependencies_manual.pdf

Hu, M., & Liu, B. (2004). Mining and summarizing customer reviews. In *Proceedings of the tenth acm sigkdd international conference on knowledge discovery and data mining* (pp. 168–177). doi: https://doi.org/10.1145/1014052.1014073

Jiang, L., Yu, M., Zhou, M., Liu, X., & Zhao, T. (2011). Target-dependent twitter sentiment classification. In *Proceedings of the 49th annual meeting of the association for computational linguistics: Human language technologies - volume 1* (pp. 151–160).

Liu, B. (2015). *Sentiment analysis: Mining opinions, sentiments and emotions.* Cambridge University Press.

Liu, B., & Zhang, L. (2012). A survey of opinion mining and sentiment analysis. In *Mining text data* (pp. 415–463). Boston, MA: Springer US. doi: https://doi.org/10.1007/978-1-4614-3223-4_13

Ory, J. C. (2000). Teaching evaluation: Past, present, and future. *New Directions for Teaching and Learning, 2000*(83), 13–18. doi: https://doi.org/10.1002/tl.8302

Taboada, M., Brooke, J., Tofiloski, M., Voll, K., & Stede, M. (2011, June). Lexicon-based methods for sentiment analysis. *Comput. Linguist., 37*(2), 267–307. doi: https://doi.org/10.1162/COLI_a_00049

Appendix: R function for aspect scoring

```
library(wordnet)
library(NLP)
library(coreNLP)
library(cleanNLP)
library(udpipe)
library(tidyverse)
library(tidytext)
library(dplyr)
```

```r
library(lubridate)
library(reshape2)
library(lmerTest)
library(lexicon)

# Using for dependency parsing
udmodel_en = udpipe_load_model(file = "english-ewt-ud-2.3-181115
                                        .udpipe")

#input: a set of comments of teaching evaluation;a list of aspects
#output: 5 aspects scores
aspect_scoring = function(comments,aspect.list,cinf=NULL){
  comments = tolower(comments)
  size = length(comments)
  annotate = udpipe_annotate(udmodel_en,x=comments,doc_id=c(1:size))
  dataframe.annotate = as.data.frame(annotate)
  negation=c('none','not','never','neither','nobody','nowhere')
  #aspect class and sentiment score for corresponding word
  ap_rs = dataframe.annotate%>%
    select(doc_id,sentence_id,token_id,lemma,head_token_id,
            dep_rel)%>%
    filter(!dep_rel=='punct')%>%
    mutate_at(c('doc_id','sentence_id','token_id','head_token_id'),
                funs(as.numeric))%>%
    rename(word=lemma)%>%
    left_join(get_sentiments('afinn'))  %>%
    left_join(aspect.list) %>%
    mutate(id=NULL,neg=if_else(word %in% negation,1,0), sentiment=NA)

  list.ap_rs=split(ap_rs,ap_rs[,1:2])
  list.ap_rs=list.ap_rs[sapply(list.ap_rs,function(x) dim(x)[1])>0]
  for (s in 1:length(list.ap_rs)){
    for (i in 1:dim(list.ap_rs[[s]])[1]){
      if (list.ap_rs[[s]][i,]$neg == 1 ){
        hid=list.ap_rs[[s]][i,]$head_token_id
        list.ap_rs[[s]][list.ap_rs[[s]]$token_id==hid,]$value =
            -list.ap_rs[[s]][list.ap_rs[[s]]$token_id==hid,]$value
      }
    }
  }
  get_aspect_sentiment=function(ls){#ls: list of annotation file
    dep_a2o=c('nsubj','obj','obl','nmod','conj','advcl','xcomp',
        'amod','acl:relcl','advmod','acl','obl:tmod','obl:npmod',
        'iobj')
    for (s in 1:length(ls)){
```

```
    for (i in 1:dim(ls[[s]])[1]){
      if (!is.na(ls[[s]][i,]$Aspect) & ls[[s]][i,]$dep_rel %in%
            dep_a2o){
        hid=ls[[s]][i,]$head_token_id
        if(hid %in% ls[[s]]$token_id){
          ls[[s]][i,]$sentiment =
              ls[[s]][ls[[s]]$token_id==hid,]$value
        }
      }
      else if (!is.na(ls[[s]][i,]$value) &
              ls[[s]][i,]$head_token_id !=0){
        hid=ls[[s]][i,]$head_token_id
        if(hid %in% ls[[s]]$token_id){
          if (!is.na(ls[[s]][ls[[s]]$token_id==hid,]$Aspect) &
              is.na(ls[[s]][ls[[s]]$token_id==hid,]$sentiment))
            ls[[s]][ls[[s]]$token_id==hid,]$sentiment
              = ls[[s]][i,]$value
        }
      }
    }
  }
  return(ls)
}
list.ap_rs.final=get_aspect_sentiment(list.ap_rs)
data.ap_rs.final=as.data.frame(bind_rows(list.ap_rs.final))
doc_aspect_sentiment=data.ap_rs.final %>%
        group_by(doc_id,Aspect) %>%
        summarise(score=sum(sentiment,na.rm=T)) %>%
        filter(!is.na(Aspect)) %>% rename(A=score)
doc_aspect_sentiment_wide=reshape(
            data.frame(doc_aspect_sentiment),
            timevar='Aspect',idvar='doc_id',direction='wide')
doc_aspect_sentiment_wide=doc_aspect_sentiment_wide %>%
        rename(comment_id=doc_id)
# the setting only for current study with 5 aspect,users can
# change it accordingly
if (is.null(cinf))
  rst=doc_aspect_sentiment_wide
else{
  final.com.abs=cinf %>% left_join(doc_aspect_sentiment_wide)
  final.pf.abs=final.com.abs %>% group_by(profid) %>%
      summarise(ncom = n(),
                      rating = round(mean(rating, na.rm=T),2),
                      A1=round(mean(A.1, na.rm=T),2),
                      A2=round(mean(A.2, na.rm=T),2),
```

```
                          A3=round(mean(A.3, na.rm=T),2),
                          A4=round(mean(A.4, na.rm=T),2),
                          A5=round(mean(A.5, na.rm=T),2))
    rst=final.pf.abs
  }
  return(rst)
}
```

Exploring Spatio-temporal Patterns of Air Quality Index Data in China

Haokun Tang[1], Yulin Xie[2], and Binbin Lu[1]

[1] School of Remote Sensing and Information Engineering, Wuhan University, Wuhan, China
binbinlu@whu.edu.cn
[2] Jiangsu Tianyi High School, Wuxi, China

Abstract. With the rapid urbanization and economy development happening in China, air pollution has been becoming a hot topic with intensive concerns. using the historical air quality index (AQI) data, this study explores the spatial-temporal distribution of AQI value to examine the distribution air pollution in China. Combined with economic and AQI data, the researches explored the spatial heterogeneity of Chinese air pollution via spatial autocorrelation analysis and Geographically Weighted Regression (GWR) technique. The air pollution data was obtained from China Air Quality Online Monitoring and Analysis Platform. The following conclusions are reached according to the results: (1) Air quality in the northern part of China is generally worse than that in the southern part. The Northeastern three province's air pollution are good. The air quality of coastal cities is better than that of inland. (2) Air quality in winter time is generally worse than that in summer time. The primary pollutant is different between summer and winter, the summer is PM10 and the winter is PM2.5. (3) Chinese air pollution has strong positive spatial autocorrelation. The aggregation pattern is high-high concentration in the north, low-low concentration in the south. (4) The GWR model fits the data set significantly better than OLR model. Fossil consumption and industrial production seem to be the main causes of air pollution.

Keywords: Air pollution · AQI · GIS · GWR.

DOI: 10.35566/isdsa2019c7

Air pollution is apparently harmful to human health and natural environment. Evidences have been frequently found between air pollutions and healthy conditions, like mortality (Dockery et al., 1993) and human body cardiopulmonary function (Chan & Yao, 2008). Meanwhile, rapid urbanization and economy development have been significantly happening within China since the reform and opening up policy was carried out. This process, however is always accompanied with an unavoidable deterioration of the ecological environment, and air pollution is an important part of it (Chan & Yao, 2008).

A number of studies on air pollution have been frequently conducted in China. Hu, Wang, Ying, and Zhang (2014) studied the correlation between

PM2.5 and PM10 in North China and Yangtze River Delta. C. Lin et al. (2015) proposed a method for estimating PM2.5 using remote sensing satellites. Fan, Wang, and Fan. (2019) calculated the economic losses of air pollution, they found the air pollution would let government waste nearly 1% of GDP money. Teng (2019) studied the relationship between air pollution and Chronic obstructive pulmonary disease and asthma. Yanting, Xingzhao, and Zhenbo (2019) used BenMAP-CE software to make air pollution health risk assessment. However, these studies did not have spatial distribution of air pollution in China. Although Yanting et al. (2019) and Xiao et al. (2017) have studied the spatial distribution of air pollution, the study lack of temporal distribution and do not found out the relationship between the economic elements and air pollution. And the studies did not focus on the local spatial characteristics.

In this study, we visualized the distributions of air pollution of China in the spatial, temporal and spatio-temporal perspectives. In this sense, we aimed at exploring the spatio-temporal patterns of air pollution happening in China. On the other hand, we preliminarily discovered the spatially varying relationships between the air pollution and potential factors via the geographically weighted regression (GWR) technique (Brunsdon, Fotheringham, & Charlton, 1996).

1   Research Area and Data

In 2012, the Chinese government issued a new Environmental Air Quality Standard (GB3095-2012), and Air Quality Index (AQI) to replace the Air Pollution Index (API) as a new air pollution assessment standard (X. Lin & Wang, 2016) .AQI is calculated individually with different pollutants, including PM2.5, PM10, sulfur dioxide, nitrogen dioxide, ozone and carbon monoxide. The greater the AQI value is, the more serious and harmful to the human body pollution is. Table 1 shows the hierarchical standards for different levels of air quality and pollutions (Yanting et al., 2019).

In this study, AQI data is collected from the China Air Quality Online Monitoring and Analysis Platform (https://www.aqistudy.cn), ranging from 2016 to 2018. In this data set, six pollutants and daily AQI values of 363 cities in China are included. All the data files are saved in the Comma-Separated Values (CSV) format. In order to study the relationship between air pollution and relative factor, we also utilized the socio-economic data integrated from the CEInet Statistics Database (available at http://db.cei.gov.cn/). The economic data include such economical attributes as GDP, national electric consumption, area of green land, areal population, industrial output, liquefied petroleum gas, natural gas, and so on. Note that only the socio-economic data of 2016 is available, so the corresponding AQI data of the same year are used in this study.

Table 1. Hierarchical standards for different levels of air quality and pollutions

Level	AQI	Air quality	Influences
1st level	0-50	Good	tabincellAir quality is satisfactory, there is no apparent air pollution, and all kinds of people can move normally.
2nd level	51-100	Moderate	A certain number of pollutants show impacts on the health of air sensitive people. A few abnormally sensitive people are recommended reduce outdoor activities.
3rd level	101-150	Mild pollution	Mildly exacerbated symptoms and irritating symptoms may appear to healthy people with sensitive constitution. Children, the elderly, and patients with heart disease and respiratory diseases are recommended to reduce long-time and intensive outdoor exercises.
4th level	151-200	Moderate pollution	Further aggravating the symptoms of susceptible people may affect the heart and respiratory system of healthy people. Patients with diseases are recommended to avoid long-time and intensive outdoor exercises, and all the outdoor sports should be reduced.
5th level	201-300	Severe pollution	Symptoms are common in healthy people. Children, elderly, patients with heart and lung diseases are recommended to stay indoors, stop outdoor sports, and all the outdoor activities should be generally reduced.
6th level	>300	Heavy pollution	There are obvious strong symptoms, some diseases appear in advance, children, the elderly and patients are recommended to stay indoors, and outdoor activities should be generally avoided.

2   Methodology

2.1   Data Preprocessing

The raw data is kinds of messy, and filled with errors or missing records. We preprocessed them with designed procedures, including removing outliers, converting coordinate system and filling missing values. Firstly, we calculated the distribution of AQI data, detected outliers via box plots and removed them accordingly. Secondly, we joined the AQI data with coordinates of each city from the Baidu map, and converted the coordinate reference system from BD-09 to WGS84 via Python scripting. Thirdly, we sorted out all the missing values and filled them with average or zero values, as the case maybe. Finally, we also calculated the primary pollutants for the following analysis in the next step.

2.2   Spatial Autocorrelation Analysis

Air is featured in flexible flowing with following airflows. Air pollution, thus could significantly affect the neighboring areas (Jiaren, Zhencheng, Yu, Peng, & Chen, 2011) and exhibit apparent spatially correlated patterns. The Moran index (Moran's I) is commonly used to measure the spatial autocorrelation (Moran, 1950). In this study, we used the Moran's I to judge the global agglomeration or discrete features of air quality in geospatial space for each city, so as to determine whether there is significant spatial agglomeration for Chinese city air qualities. The formula of Moran's I could be expressed as follows,

$$I = \frac{n \sum_{i=1}^{n} \sum_{j=1}^{n} w_{ij} z_i z_j}{s_0 \sum_{j=1}^{n} z_j^2} \tag{1}$$

where z_i is the attribute deviation of element i from its mean value,i.e. $(x_i - \bar{X})$, w_{ij} is the spatial weight between element i and j, n is the number of elements, and s_0 is the aggregation of all the spatial weights. The value of Moran's I ranges from -1 to 1. When the Moran's I is close to 1, it means the air pollution is positively agglomerated, i.e. exhibiting high-high or low-low aggregations; when Moran's I is close to -1, it means the air pollution is negatively correlated, i.e. exhibiting high-low or low-high aggregations; when Moran's I is close to 0, it means that the air pollution distribution within China is random and irregular. In this study, the targeting element is the AQI value of each city.

Furthermore, we also use the local indicators of spatial association (LISA) to measure the local autocorrelations of air pollutions among the cities (Anselin, 1995). In this study, we used local Moran's I to reflect the LISA features, and its formula is as follows,

$$I_i = \frac{x_i - \bar{X}}{S_i^2} \sum_{j=1, j \neq i}^{n} w_{i,j} \left(x_j - \bar{X} \right) \tag{2}$$

where I_i is local Moran's I value at element i,x_i is attribute of element i,$\bar{X}$ is average of this attribute,w_{ij} is the spatial weight between element i and j, S_i^2 formula is as follows,

$$S_i^2 = \frac{\sum_{j=1,j\neq i}^{n} \left(x_j - \bar{X}\right)^2}{n-1} \tag{3}$$

n is the number of elements. Different from global Moran's I, LISA present the local autocorrelations. When the local Moran's I is close to 1 or -1, it means the air pollution is positively agglomerated or negatively agglomerated at local, not the whole study region.

2.3 Geographically Weighted Regression

Geographically Weighted Regression (GWR) is local modelling technique to estimate regression models with spatially varying relationships. Compared with a basic linear regression model, the coefficients in a GWR model are functions of spatial locations, attached to the spatial coordinates of the locations(Brunsdon et al., 1996). A general form of basic GWR models at each regression point (u_i, v_i) could be expressed as,

$$y_i = \beta_0 (u_i, v_i) + \sum_{k=1}^{P} \beta_k (u_i, v_i) x_{ik} + \varepsilon_i \quad i = 1, 2 \cdots n \tag{4}$$

where y_i is the dependent variable at location i, x_{ik} is the value of the kth explanatory variable at location i, m is the number of explanatory variables,$\beta_0 (u_i, v_i)$ is the intercept parameter at location i, $\beta_o (u_i, v_i)$ is the local regression coefficient for the kth explanatory variable at location i, are the coordinate of location i, and ε_i is the random error at location i.

GWR makes a point-wise calibration concerning a 'bump of influence': around each regression point nearer observations have more influence in estimating the local set of coefficients than observations farther away (Fotheringham, Charlton, & Brunsdon, 1998). In essence, GWR measures the inherent relationships around each regression point i with fitting each set of regression coefficients by weighted least squares. The matrix expression for its calibration is shown as formula (5).

$$\hat{\beta} (u_i, v_i) = \left(X^T W (u_i, v_i) X\right)^{-1} X^T W (u_i, v_i) y \tag{5}$$

where X is the matrix of the explanatory variables with a column of 1s for the intercept, y is the vector of the dependent variable, $\hat{\beta} (u_i, v_i) = (\beta_0 (u_i, v_i), \cdots,$ $\beta_n (u_i, v_i))^T$ is the vector of n+1 local regression coefficients, $W (u_i, v_i)$ is the diagonal matrix denoting the geographical weighting of each observed data for regression point i, which is defined as:

$$W_i = \begin{bmatrix} w_{i1} & 0 & \cdots\cdots & 0 \\ 0 & w_{i2} & \cdots\cdots & 0 \\ \cdot & \cdot & \cdot & \cdot \\ \cdot & \cdot & \cdot & \cdot \\ 0 & 0 & \cdot & \cdot & w_{in} \end{bmatrix} \tag{6}$$

Here,the weighting scheme $W(u_i, v_i)$ is calculated with a kernel function based on the proximities between regression point i and the n data points around it. In this study, we used the Bisquare kernel function to calculate the distance-decaying spatial weights w_{ij}, which could be expressed as follows,

$$w_{ij} = \begin{cases} \left[1 - \left(\frac{d_{ij}}{b}\right)^2\right]^2 & d_{ij} \leq b \\ 0 & d_{ij} \geq b \end{cases} \tag{7}$$

where d_{ij} is the spatial distance between location i and j, b is bandwidth. Generally speaking, the parameter bandwidth determines how many observations are concerned locally for each location-wise calibration.

In order to judge whether the GWR technique significantly works better rather than using the OLS model. A statistical hypothesis test could be conducted beforehand when using the GWR technique(Leung, Mei, & Zhang, 2000). There are three kinds of F tests with a common null hypothesis, i.e. 'there is no significant difference between OLS and GWR models for the given data set'.

3   Results

3.1   Spatial distribution analysis

We used the mapping scatterplot to show the air pollution spatial distribution of China in 2017 and 2018, as shown in Fig. 1 and Fig. 2. All the points present the air quality for each city. The bigger and darker color the point appears, the worse the air quality corresponds. On the whole, the conditions of air pollution in northern China are worse than southern China. The most serious air pollutions happened in Beijing-Tianjin-Hebei region, Shanxi province and Xinjiang province. For the Shanxi province, the development of coal industry could be the principle cause of its bad air qualities. Dry and windy climate, intensive habitat and heavy industries could be the main reasons for the serious air pollutions in the Beijing-Tianjin-Hebei region. In contrast, the Central China, Shanghai municipality, and Sichuan province are less polluted, South China, Northeast China, Tibet, Yunnan and other places have the best air quality.

3.2   Temporal distribution analysis

Fig. 3 shows the monthly AQI change in 2016-2018. The y-axis is AQI value, the x-axis is timeline by month. We used rectangles in three different colors to show the different features. The red rectangle means the air quality is worst. The purple rectangle shows the air pollution could be bad in April and June of each year. The green rectangle indicates the air qualities in summer time are the best. It shows the Chinese air quality is obviously affected by seasonal climate changes. From 2016 to 2018, air pollution always shows similar trends: from January to March each year, the AQI value drops sharply, and it rises slightly from April to June as purple rectangle points out. From green rectangles, we can

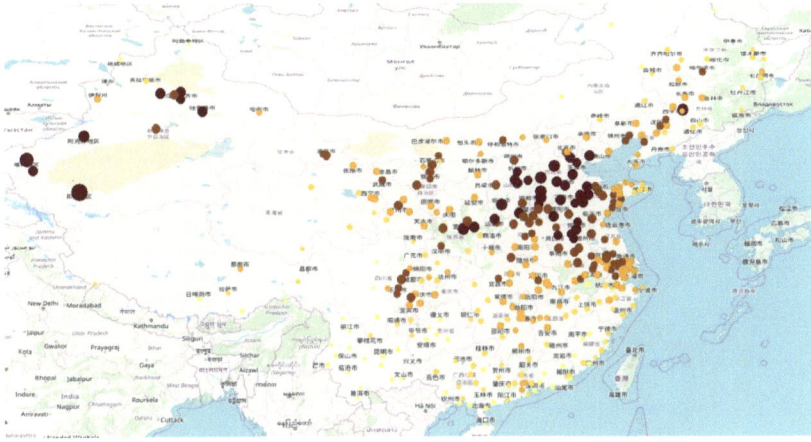

Fig. 1. Air pollution spatial distribution of China in 2017

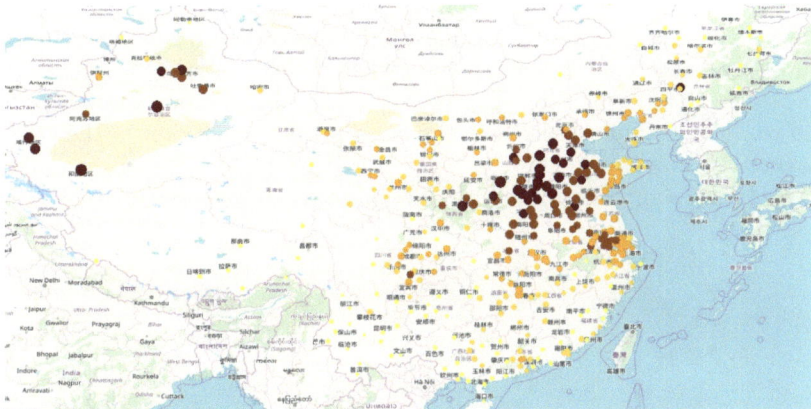

Fig. 2. Air pollution spatial distribution of China in 2018

see summer is the best year for air quality, and since October, the air quality will deteriorate dramatically until the January as red rectangle shows. January is the worst month for air quality. It also shows that the air qualities are better and better year by year, saying the AQI values in 2017 is better than those in 2016, and 2018 is better than 2017.

There is another interesting thing we need pay attention to, that the AQI will rise slightly from April to June. It's obvious in 2017 and 2018 but 2016 is not.

Fig. 4 shows the monthly primary pollutant trend from 2016 to 2018 in China. The x-axis is timeline. The y-axis presents the number of pollutants which become the primary pollutants per month, the higher means this kind of pollutant is more likely become the primary pollutant in this month. The

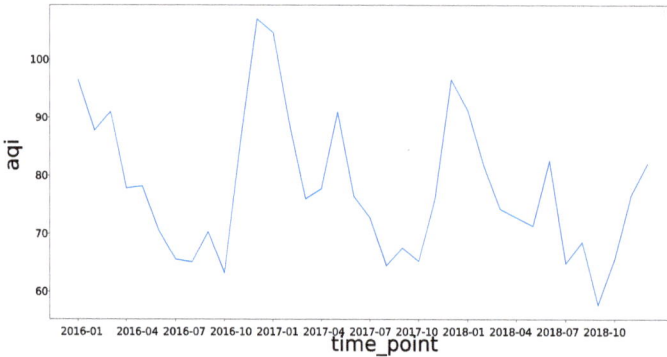

Fig. 3. Monthly average AQI from 2016 to 2018

primary pollutants in winter in China are mainly PM$_{2.5}$, however in summer, PM10 becomes the main pollutant. And when the PM$_{2.5}$ rise, the PM$_{10}$ down. They are absolutely opposite. The change of O$_3$ has obvious regularity, from March to June, O$_3$ quickly become the main primary pollutant in the country. it because the sunshine in summer is stronger than winter, it will speed up the nitrogen oxide in air occur chemical reaction to produce O$_3$.

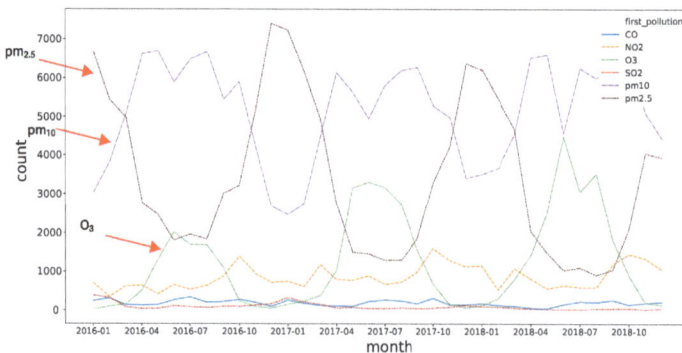

Fig. 4. Monthly primary pollutant trend from 2016 to 2018 in China

For the other pollutants, the NO$_2$ is much higher than the SO$_2$ and CO. Generally speaking, the NO$_2$ shows a peak in October every year. There is slight CO pollution for every month.

3.3   Spatial-temporal distribution

It is obvious that the distribution of primary pollutants in China is so different spatially and temporally. Fig. 5 shows the primary pollutants for each month in 2018. Each point presents a city, the different colors mean the different kind of pollutant become the primary pollutant in this month. As the Fig. 5 show, the red point is NO2 pollution point, it means the NO_2 is the primary pollutants in the month. According to the purple rectangle show in the Fig. 5, the NO_2 pollution in China is mainly occurred in Shanghai and Guangdong Province. These two areas are always let the NO_2 become the primary pollutant. It because the car is the mainly NO_2 source in city. These two cities are the most developed city in China, these two cities have so many cars.

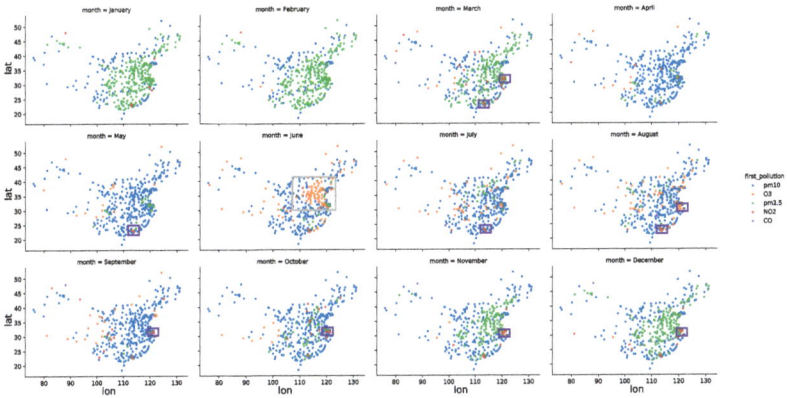

Fig. 5. 2018 Primary pollutants in cities each month

According to the Fig. 5, another conclusion we can make is that the O_3 pollution occurred in Northern of China at June. We make the same conclusion according to Fig. 4, but we can see the spatial distribution in Fig. 5, As the gray rectangle indicating, the orange point is O_3 pollution point, and it occurs in Northern of China at June aggregation. It means the Northern of China will take a suffer in O_3 in June. It happened suddenly and disappeared quickly, it means this phenomenon is happened at a special time point and a special area.

Fig. 6 shows the monthly AQI values for each city in 2018. Each point presents the city and the color indicates the AQI value. Similar as the temporal analysis, the air quality is not good in winter. As Fig. 6 show, the Northern China tends to be the worst area. The northern climate is dry, the vegetation coverage is low, the terrain is low, the convection is not easy to form, and the diffusion conditions are poor. The particulate matter carried by the dust storm stays in the air for a long time. The air quality is generally good in summer, and the air quality is good in most areas except Xinjiang province. The coastal areas of Yunnan, Guangxi and Guangdong provinces and Hainan Province are affected by the southwest monsoon, the atmosphere vertical movement is active, the water vapor transportation is good, and the rain is rainy, windy weather is conducive to the dissipation of air pollutants, so the air quality is excellent. In Hetian, Xinjiang, due to sparse vegetation and frequent sandstorms, air pollution is still serious in summer.

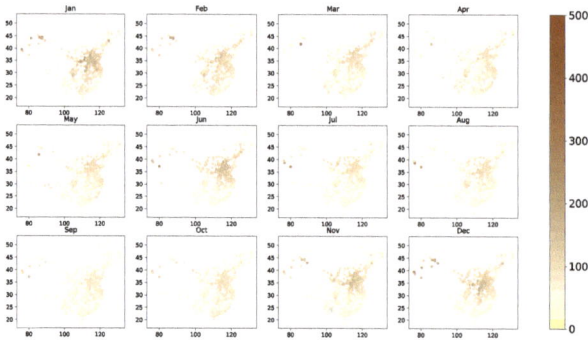

Fig. 6. AQI monthly index of each city in 2018

In order to find out whether the different life behavior in weekday and weekend influence the air quality. According to some studies Shuoben, Lei, Shuliang, Yechao, and Athanase (2018); Zhong and Sai-xing (2019), there are many differences in people's lifestyle between weekday and weekend such as transformation mode and entertainment mode. Fig. 7 is a spatial-temporal cubic shows the AQI value in China from Monday to Sunday. Every bar is composed of seven small bars, the bottom of small bar means the average Monday air pollution, the top is Sunday. As it shows, there is no obvious difference between weekday and weekend, so the people's lifestyle doesn't affect the air quality.

3.4   Spatial heterogeneity analysis

Fig. 8 is global Moran's I scatter plot of AQI, according to it, the Moran's I is 0.75. Most of points are in the first and third quadrants. It means the spatial distribution in China air pollution has strong positive autocorrelation.

Fig. 9 is LISA cluster scatter plot of AQI. From this picture we can see the Chinese air pollution has two kind of aggregation. The HH aggregation mainly happened in Northern China like Beijing, Tianjin, Hebei province, Shanxi province, Shandong Province, Henan province and some areas in Xinjiang province. It means the air quality is bad in that regions. The LL aggregation mainly happened in southern coastal provinces and Xizang province, Qinghai province and some areas in three provinces in Northeastern of China, because these cities is developing, and some cities have higher altitudes.

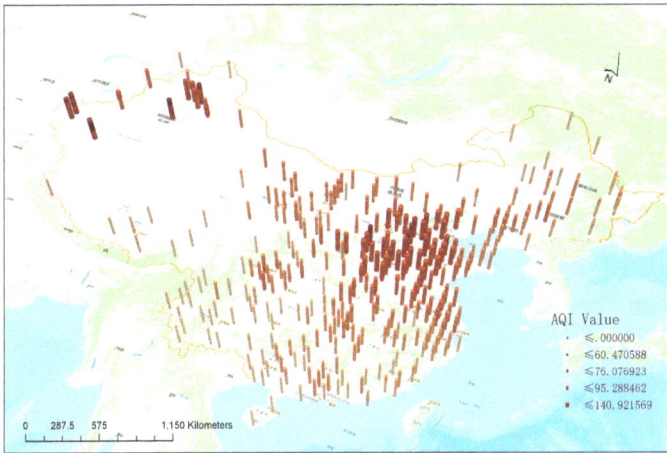

Fig. 7. Monday to Sunday AQI in China

Fig. 10 is the result of Geographically weighted regression model selection. This procession can select the attribute which should participate in calculation. In this procession, our target is to decrease the CV value to find the most important attributes. As it shows in picture, there is no sudden decrease after the thirtieth model number, so we can use the model around thirtieth number which the red arrow point to. Table 2 is attributes been selected.

Table 2. Selected attributes

Fields	describe
na_gas_f	Total domestic gas supply for urban artificial gas and natural gas
pet_gas_f	Total domestic gas supply of municipal LPG
ind_ov_n	Gross industrial output value of domestic enterprises above designated size (5 million CNY)

The natural gas, artificial gas, LPG and the industry are well known the influence factors of air pollution. The results of model selection also prove it.

Table 3 shows the result of regression, there are two methods, the R-square values are calculated by the GWR model (R-square value: 0.72) is much higher than the OLR model (R-square value: 0.11). It means GWR model fit the data set much better than the OLR model. The adjusted R square value of GWR

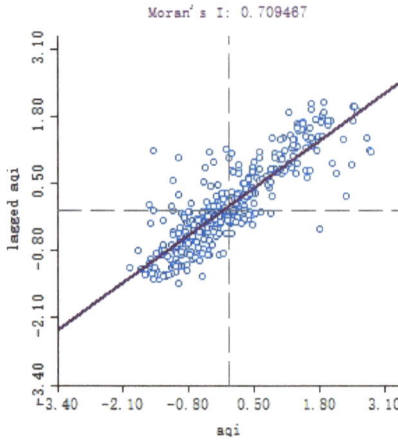

Fig. 8. Moran's I scatter plot of AQI

method (R-square value: 0.68) shows the positive relationship between selected attribute and air pollution. It indicated the amount of the natural gas, artificial gas, LPG and the industry output may be the import influence factors of air pollution.

Table 3. Selected attributes

Method	R-square	Adjusted-R-square
OLS	0.11	0.09
GWR	0.72	0.68

F test is used to find out whether the relationship between AQI and economic attribute has spatial heterogeneity. The result of F_1 test and F_2 test are 2.2e-16, the alpha level is below the 0.001, it means the probability of collecting the observed sample is small under the null hypothesis. There is more than 99.9% probability to reject the null hypothesis, so there is significant difference between OLR and GWR models for the AQI and economic data. The result of F_3 test is shown on Table 4.

Table 4. Selected attributes

Fields	F3 test value	alpha level
Intercept	2.2e-16	<0.001
na_gas_f	0.046	0.01
pet_gas_f	5.3e-08	<0.001
ind_ov_n	1.3e-08	<0.001

All the attributes are more than 99% to reject the null hypothesis, so parameter β_k is different over the study region.

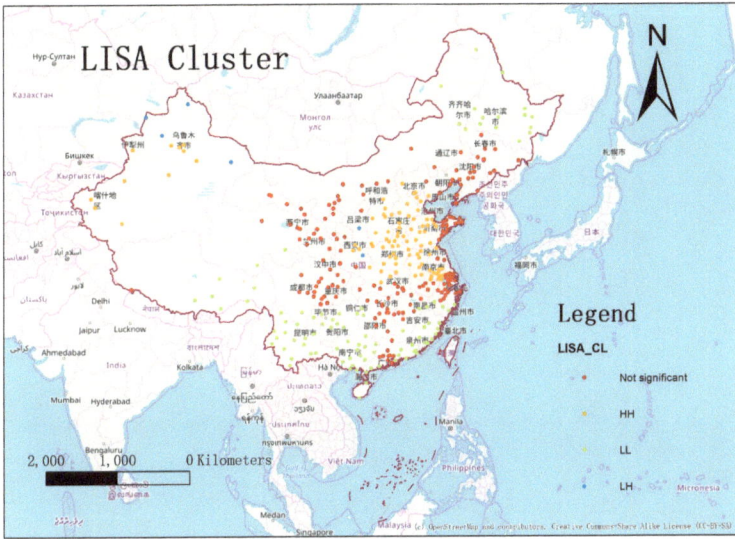

Fig. 9. LISA cluster scatter plot of AQI

4   Conclusion and Future work

This study aims to provide a qualitative exploration of the spatio-temporal patterns of air pollution in China via spatio-temporal visualizations with AQI data. We explored the spatial heterogeneities in the data relationships between air pollution and relative factors, globally and locally via spatial autocorrelation analysis and the GWR technique. With the results, we could conclude this work as the following points.

(1) Air quality in the northern part of China is generally worse than that in the southern part, except the Northeastern China.

(2) Air quality in winter time is generally worse than that in summer time. PM2.5 tends to be the primary pollutant in winter and PM10 appears as the primary pollutant in summer.

(3) Air pollution has strong positive autocorrelation. High-high concentration is mainly happened in north and low-low concentration is mainly happened in south.

(4) Using fossil and industrial production seem to be the main influence factors of air pollution.

In this study, we explored the spatio-temporal patterns of air pollutions within China, specifically for the main cities with AQI data. How to make effective predictions, however is a topic to be studied in the future. Furthermore, this study only take economical factors into consideration, the natural factors are also affect the air quality, we will consider it in the future.

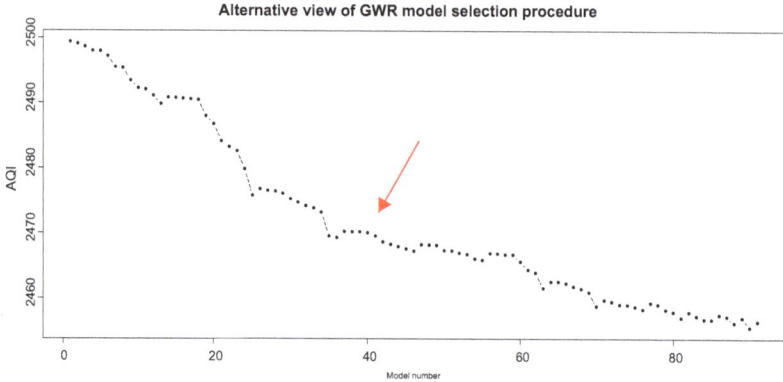

Fig. 10. Model Selection

References

Anselin, L. (1995). Local indicators of spatial association-lisa. *Geographical analysis*, *27*(2), 93–115. doi: https://doi.org/10.1111/j.1538-4632.1995.tb00338.x

Brunsdon, C., Fotheringham, A. S., & Charlton, M. E. (1996). Geographically weighted regression: a method for exploring spatial nonstationarity. *Geographical analysis*, *28*(4), 281–298. doi: https://doi.org/10.1111/j.1538-4632.1996.tb00936.x

Chan, C. K., & Yao, X. (2008). Air pollution in mega cities in china. *Atmospheric environment*, *42*(1), 1–42. doi: https://doi.org/10.1016/j.atmosenv.2007.09.003

Dockery, D. W., Pope, C. A., Xu, X., Spengler, J. D., Ware, J. H., Fay, M. E., ... Speizer, F. E. (1993). An association between air pollution and mortality in six us cities. *New England journal of medicine*, *329*(24), 1753–1759. doi: https://doi.org/10.1056/NEJM199312093292401

Fan, F., Wang, H., & Fan., L. (2019). Health damage and economic loss assessment of air pollution in jing-jin-ji region. *Ecological Economy*(09), 157-163.

Fotheringham, A. S., Charlton, M. E., & Brunsdon, C. (1998). Geographically weighted regression: a natural evolution of the expansion method for spatial data analysis. *Environment and planning A, 30*(11), 1905–1927. doi: https://doi.org/10.1068/a301905

Hu, J., Wang, Y., Ying, Q., & Zhang, H. (2014). Spatial and temporal variability of pm2. 5 and pm10 over the north china plain and the yangtze river delta, china. *Atmospheric Environment, 95*, 598–609. doi: https://doi.org/10.1016/j.atmosenv.2014.07.019

Jiaren, S., Zhencheng, X., Yu, L., Peng, X., & Chen, L. (2011). Advances in the effect of climate change on air quality. *Climatic and Environmental Research (in Chinese), 16*(6), 805–814.

Leung, Y., Mei, C.-L., & Zhang, W.-X. (2000). Statistical tests for spatial non-stationarity based on the geographically weighted regression model. *Environment and Planning A, 32*(1), 9–32. doi: https://doi.org/10.1068/a3162

Lin, C., Li, Y., Yuan, Z., Lau, A. K., Li, C., & Fung, J. C. (2015). Using satellite remote sensing data to estimate the high-resolution distribution of ground-level pm2. 5. *Remote Sensing of Environment, 156*, 117–128. doi: https://doi.org/10.1016/j.rse.2014.09.015

Lin, X., & Wang, D. (2016). Spatio-temporal variations and socio-economic driving forces of air quality in chinese cities. *Acta Geographica Sinica, 71*(8), 1357–1371. doi: https://doi.org/10.11821/dlxb201608006

Moran, P. A. (1950). Notes on continuous stochastic phenomena. *Biometrika, 37*(1/2), 17–23. doi: https://doi.org/10.2307/2332142

Shuoben, B., Lei, W., Shuliang, Y., Yechao, Y., & Athanase, N. (2018). Spatial-temporal distribution of the pick-up and drop-off of taxi passengers in nanjing based on gps data. *China Sciencepaper, 13*(09), 1023-1028.

Teng, J. (2019). Short-term effects of hospitalization for outdoor air pollution and respiratory diseases. *master, University of Science and Technology of China*. (Available from CNKI)

Xiao, Y., Tian, Y., Wenxuan, X. U., Jingjing, W. U., Tian, L., & Liu, J. (2017). Spatiotemporal pattern changes of air quality in china from 2005 to 2015. *Ecology & Environmental Sciences, 26*(2), 243-252.

Yanting, X., Xingzhao, L., & Zhenbo, W. (2019). Spatial-temporal distribution characteristics of air quality in chinese cities based on the aqi index. *Journal of Guangxi Normal University(Natural Science Edition), 37*(1), 187-196.

Zhong, H., & Sai-xing, H. (2019). Study on leisure and entertainment time of residents on weekdays and rest days. *Economic Forum*(06), 139-145.

VBTree Tutorial: Automatic Completion of Group Operation on Structural Dataset

Chen Zhang[1], Huakang Bian[2], Kenta Yamanaka[2], and
Akihiko Chiba[2]

[1] Department of Materials Processing, Graduate School of Engineering, Tohoku
University, Sendai, JAPAN
`chenzhang2013@imr.tohoku.ac.jp`
[2] Institute for Materials Research, Tohoku University, Sendai, JAPAN

Abstract. This paper introduces an R package designed for engineering
practitioners and researchers, which permits that most of the plotting
and data analyzing tasks could be directly achieved from their summary
table of dataset. It is even unnecessary for users to master some complicated data reshaping operations in R aforehand. By understanding
the built-in managing logic of VBTree as well as essential knowledge of
R, even a novice can make their data processing workflow be achieved
automatically.

Keywords: R package · VBTree

DOI: 10.35566/isdsa2019c8

1   Introduction

For most engineering practitioners and researchers, their data are commonly the
exported results of the built-in application in their apparatus. For most experimental designs, they should talk about how the dependent variables respond to
different factors. After a series of experiments using different factors combinations, a tedious but essential task is to finish the data processing tasks. These
tasks generally consist of some simple data visualization at different levels of the
identical factor, capturing the linear relationship between the dependent variable and some specific factors, etc. However, with the increase in experimental
factors, the time cost on the data processing tasks will increase accordingly.
Athough these tasks can be fulfilled using programming languages, the reshaping operation for the dataset is required beforehand, which, in turn, increases
the learning cost for engineering practitioners and researchers. VBTree is an R
package aimed to solve these problems. By importing VBTree in R script, most
of the data reshaping steps can be skipped, and then data analysts can directly
build their codes based on the original dataset.

2   Essential Concepts

To perform group operation on an original dataset correctly, the data organization logic in VBTree should be understood. To illustrate why VBTree can identify

the experimental variables precisely from the summary table, the following sub-sections will take the `ToothGrowth` dataset as an instance, through which the group processing on a dataset can be achieved without reshaping operations.

2.1   Summary Table in VBTree Style

`ToothGrowth` is the dataset with information on the length of odontoblasts (cells responsible for the tooth growth) in 60 guinea pigs (Bliss, 1952), using different two supplements (vitamin C and orange juice, denoted as VC and OJ respectively) and three doses (0.5, 1.0 and 2.0 milligram per day). In R, it has already been arranged in the following format.

```
> head(ToothGrowth)
   len supp dose
1  4.2   VC  0.5
2 11.5   VC  0.5
3  7.3   VC  0.5
4  5.8   VC  0.5
5  6.4   VC  0.5
6 10.0   VC  0.5
```

Nevertheless, this format is not what most lab technicians would do empirically, since nobody will repeatedly note down the experimental conditions for each observation. In general, they record the all observations under the certain experimental conditions then just write down that conditions only once.

However, by importing the `VBTree` package, it allows our codes using data frame with structural column names sorted by different condition combinations, whose arrangement is largely consistent with our conventional summary table. Take `ToothGrowth` as the example, by executing the following codes:

```
TGvbt <- as.data.frame(ToothGrowth)
TGvbt <- cbind(TGvbt[1:10,], TGvbt[11:20,], TGvbt[21:30,],
TGvbt[31:40,], TGvbt[41:50,], TGvbt[51:60,])
TGvbt <- TGvbt[,c(1:6)*3-2]
colnames(TGvbt) <- c("len-VC-0.5", "len-VC-1", "len-VC-2",
"len-OJ-0.5", "len-OJ-1", "len-OJ-2")
```

The dataset will be reorganized with the following arrangement:

```
> head(TGvbt)
  len-VC-0.5 len-VC-1 len-VC-2 len-OJ-0.5 len-OJ-1 len-OJ-2
1        4.2     16.5     23.6       15.2     19.7     25.5
2       11.5     16.5     18.5       21.5     23.3     26.4
3        7.3     15.2     33.9       17.6     23.6     22.4
4        5.8     17.3     25.5        9.7     26.4     24.5
5        6.4     22.5     26.4       14.5     20.0     24.8
6       10.0     17.3     32.5       10.0     25.2     30.9
```

It is obvious that elements in this **VBTree** style data frame are all observation data (numeric ones), and all experimental conditions are noted in each column names. Different types of experimental conditions are separated by the hyphen symbols. Users can utilize any application they are familiar with, to build such a summary table for their experiment data. After this summary table is imported in **R** successfully, users can directly initiate the coding step for data processing.

2.2   Data Management Logic in VBTree

Instead of directly operating on data itself, **VBTree** achieves data management by manipulating strictly-defined column names. Take the **ToothGrowth** case as an example. After we reorganized its structure, all experimental observations (length) are arranged with the structure of "Supplement-Dose" combination column-wisely. For programming language itself, it is impossible to identify the intrinsic attribute of our experimental observations. As an example, if we want the subset data of vitamin C supplement with the dose of 2.0 milligram per day, it is not a big deal for humans. However, it is very difficult to tell the program that is the 3rd column in **VBTree** style data frame since the column name "len-VC-2" is merely a meaningless string character object.

From the **VBTree** data frame of **ToothGrowth**, it is not difficult to find the data consists of one type of observation (length), two kinds of supplements (VC and OJ) and three different dose (0.5, 1 and 2 mg/d). We can make indices for these three factors respectively. Thus, it is likely to index all column names (string character objects) as a specific tensor (or array). By calling the methods of tensor or array using specific indices, we can extract the name(s) of desired columns, through which the desired subset in the **VBTree** data frame could be visited indirectly.

Following the terminology in the experimental design method, the different types of the experimental variables are called factors (or layers), and the different values in the same type of variables are termed as levels. When we build a tensor or array containing all column names using **VBTree**, the built-in index is strongly related to the intermediate data structure called vector binary tree, which is also the source of the package's name. By running the following codes:

```
library(VBTree)
arr <- dl2arr(chrvec2dl(colnames(TGvbt)))
```

all column names of **VBTree** style **TGvbt** data frame are allocated into the **arr** array. To visualize its corresponding vector binary tree:

```
> arr2vbt(arr)
$tree
$tree[[1]]
[1] "len"

$tree[[2]]
$tree[[2]][[1]]
```

```
[1] "OJ" "VC"

$tree[[2]][[2]]
$tree[[2]][[2]][[1]]
[1] "0.5" "1"    "2"

$tree[[2]][[2]][[2]]
list()

$dims
[1] 1 2 3

attr(,"class")
[1] "Vector.Binary.Tree"
```

It is noteworthy that the levels' order in the 2nd layer in the variable arr (arr $\in \mathbb{N}^{+1 \times 2 \times 3}$) has been sorted alphabetically, which is intuitively different from the arrangement of original data frame. In such scenario, the mapping relationship between all column names and the corresponding indices has been developed successfully. Instead of `subset(ToothGrowth, supp=="VC" & dose=="0.5")`, `TGvbt[,arr[,2,1]]` can be simply used to specify the subset under the same experimental conditions:

```
> TGvbt[,arr[,2,1]]
[1]   4.2 11.5   7.3   5.8   6.4 10.0 11.2 11.2   5.2   7.0
```

More importantly, if we have to do some specific group operations on the ToothGrowth data, the expression `TGvbt[,arr[,i,j]]` will be more readable when allocated in the loop structure, where the i and j are both the loop indicators.

3   Group Operation Methods through VBTree

After understanding the essential concepts of the VBTree package, for a specific VBTree data frame, the grouped data processing becomes quite easy. By flexibly calling the methods in the R data frame using the mapping relationship between the original data frame and the corresponding column names' tensor (array) established by VBTree, it is easy to implement data operations, e.g., subset extraction and marginalization calculation under specific conditions. Following the mentioned instance, this section illustrates those group operations on data using specific examples, and then the advantages of using VBTree package are discussed.

3.1   Group Operation Using Column Names Tensor

According to our previous demonstration, all column names of TGvbt have been allocated in a specific array (or tensor) arr. Since the indexing relationship has already been developed, users can use TGvbt[,arr[,2,1]] to specify the data using vitamin C supplement with 0.5 mg/d dose, or perform TGvbt[,arr[,2,]] to obtain all vitamin C supplied group:

```
> TGvbt[,arr[,2,]]
    len-VC-0.5 len-VC-1 len-VC-2
1          4.2     16.5     23.6
2         11.5     16.5     18.5
3          7.3     15.2     33.9
4          5.8     17.3     25.5
5          6.4     22.5     26.4
6         10.0     17.3     32.5
7         11.2     13.6     26.7
8         11.2     14.5     21.5
9          5.2     18.8     23.3
10         7.0     15.5     29.5
```

3.2   Group Operation Using Index Vector

To remove the dependence between function body and selected factor, the method through index vector should be built. VBTree can build a subtree from original tree, using a proper visiting vector (see the documentation of the function vbtsub()). The value −1 is specified to control executing traversal on all levels of current factor in vbtsub(). Thus, by compiling the following function:

```
indexing <- function(x, index){
  vbt <- dl2vbt(chrvec2dl(colnames(x)))
  subvbt <- vbtsub(vbt, index)
  result <- as.vector(unlist(vbt2arr(subvbt)))
  return(result)
}
```

The index vector $c(-1, 2, -1)$ can be used to specify all vitamin C supplement-related column names:

```
> indexing(TGvbt, c(-1,2,-1))
[1] "len-VC-0.5" "len-VC-1"   "len-VC-2"
```

As the result, the previous example subset can identically be extracted by:

```
> TGvbt[,indexing(TGvbt, c(-1,2,-1))]
    len-VC-0.5 len-VC-1 len-VC-2
1          4.2     16.5     23.6
2         11.5     16.5     18.5
3          7.3     15.2     33.9
4          5.8     17.3     25.5
5          6.4     22.5     26.4
6         10.0     17.3     32.5
```

7	11.2	13.6	26.7
8	11.2	14.5	21.5
9	5.2	18.8	23.3
10	7.0	15.5	29.5

This method is virtually to use a real vector object to control the group operations on dataset. If a controlling binary vector `select` is defined as the argument,
of which the 1st and 2nd elements represent the fixed layer and the fixed level
at this layer, respectively, the following function can be generically developed to
select the group on the subset without any modification on the function body:

```
GroupSlt <- function(x, select){
    dims <- dl2vbt(chrvec2dl(colnames(x)))[[2]]
    idx <- rep(-1, length(dims))
    idx[select[1]] <- select[2]
    result <- x[,indexing(x, idx)]
    return(result)
}
```

Subsequently, the group operations are easy to apply.

3.3 Group Operation Traversed on Dataset Using Specific Condition

The previous demonstration of `GroupSlt` by specifying the rank of desired layer
can implement the traversal operation on dataset. The respective numbers of
levels in all layers can be yielded from the right subtree of the original vector
binary tree:

```
> arr2vbt(arr)[[2]]
[1] 1 2 3
```

Accordingly, some group-selection-dependent operations, e.g. marginal statistics
on specific factor, can be implemented using `VBTree` effortlessly. For instance,
to output the mean and standard deviation of a selected subset, the following
function can be defined:

```
msdprint <- function(x, fac){
    vbt <- dl2vbt(chrvec2dl(colnames(x)))
    dims <- vbt[[2]]
    dl <- vbt2dl(vbt)
    rpt <- dims[fac]
    for (i in 1:rpt) {
        cat(dl[[fac]][i], ":\n")
        sltsubset <- GroupSlt(x, c(fac, i))
        temp <- as.vector(unlist(sltsubset))
        cat(paste("mean ", round(mean(temp), 3), "\n"))
        cat(paste("sd ", round(sd(temp), 3), "\n"))
```

```
    }
}
```

where `fac` denotes the argument (a positive integer) to specify the fixed layer. By setting the `fac` as 2, the function can simple output the values of mean and standard deviation of vitamin C and orange juice supplement, respectively:

```
> msdprint(TGvbt, 2)
OJ :
mean   20.663
sd   6.606
VC :
mean   16.963
sd   8.266
```

Likewise, the statistics for the variable dose can be obtained by setting the `fac` as 3.

3.4 Generic Group Operation Solution Designed for Customized Traversal

The previous examples have two very serious flaws. The first is that each assignment of a fixed layer for controlling the traversal algorithm includes one factor only, whereas the term "group" can also be defined as some specific combinations of multiple factors.

The other flaw is that previous demonstrations only work correctly on a single data type. In other words, all our values in the dataset are the "length" of odontoblasts. This situation is not common, especially in the summary table. For instance, the dataset `iris3` has two types of numeric measurements, namely width and length (Becker, Chambers, & Wilks, 1988), and the dataset *Orange* includes the age of tree and circumferences simultaneously (Draper & Smith, 2014).

It helps to reconsider the structure of column names of **VBTree** data frame TGvbt: "len-supp-dose" is actually "data-condition1-condition2". The data layer only has one level "length". If there are multiple levels in a data layer, the location of the data layer should be declared in advance. In the following, a pseudo dataset is expanded from `ToothGrowth` that also contains the results of the surface area and cell weight of odontoblasts. Except for the supplement and the dose, two new factorial variables are expanded: the supplementation time (morning, afternoon and evening) and nutrition companies who offer the supplements (A and B):

```
list(c("len", "area", "weight"),
c("VC", "OJ"),
c(0.5, 1, 2),
c("morning", "afternoon", "evening"),
c("A", "B")) -> psTGd1
```

```
class(psTGdl) <- "Double.List"

colnames_vec <- as.vector(unlist(dl2arr(psTGdl)))
exTGvbt <- matrix(rnorm(length(colnames_vec)), nrow = 1)
colnames(exTGvbt) <- colnames_vec
exTGvbt <- as.data.frame(exTGvbt)
```

The expanded summary table (exTGvbt) has $3 \times 2 \times 3 \times 3 \times 2 = 108$ columns. The data layer is of the first factor. To implement the generic group operations, it is necessary to build a 3 dimentional tensor to rearrange all column names. To achieve better visualization in R, the 2nd and 3rd dimensions of that tensor are defined to arrange the select factor(s) and the levels in data layer, respectively. By setting two arguments dataly (integer) and grp (integer vector) to specify the data layer and the combinations grouped by fixed factor(s), the tensor for rearranged column names can be uniquely determined. The example codes are shown below:

```
require(tensorA)
grparr <- function(x, dataly, grp){
  arr <- dl2arr(chrvec2dl(colnames(x)))
  vbt <- arr2vbt(arr)
  dims <- vbt[[2]]
  subst <- pos.tensor(dims[grp])
  if(length(subst[,1])!=prod(dims[grp])){
    subst <- matrix(1, prod(dims[grp]), length(dims[grp]))
    subst[,length(dims[grp])] <- c(1:prod(dims[grp]))
  }
  rpt <- length(subst[,1])
  margin <- prod(dims[-dataly])/prod(dims[grp])
  ids <- matrix(-1, rpt, length(dims))
  result <- array(NA, c(margin, rpt, dims[dataly]))
  for (k in 1:dims[dataly]) {
    ids[,dataly] <- rep(k, rpt)
    for (i in 1:rpt) {
      ids[i,][grp] <- subst[i,]
    }
    result1 <- matrix(NA, length(ids[,1]), margin)
    for (i in 1:length(ids[,1])) {
      temp <- as.vector(unlist(vbt2arr(vbtsub(vbt, ids[i,]))))
      result1[i,] <- temp
    }
    result[,,k] <- t(result1)
  }
  return(result)
}
```

Using the calculation for supplement by setting the `dataly` as 1 while the `grp` as 2, the original column names' tensor with the dimension of $\mathbb{N}^{+3\times2\times3\times3\times2}$ can be rearranged into a 3D tensor as shown in the following results:

```
> grparr(exTGvbt, 1, c(2))
, , 1
```

	[,1]	[,2]
[1,]	"area-OJ-0.5-afternoon-A"	"area-VC-0.5-afternoon-A"
[2,]	"area-OJ-1-afternoon-A"	"area-VC-1-afternoon-A"
[3,]	"area-OJ-2-afternoon-A"	"area-VC-2-afternoon-A"
[4,]	"area-OJ-0.5-evening-A"	"area-VC-0.5-evening-A"
[5,]	"area-OJ-1-evening-A"	"area-VC-1-evening-A"
[6,]	"area-OJ-2-evening-A"	"area-VC-2-evening-A"
[7,]	"area-OJ-0.5-morning-A"	"area-VC-0.5-morning-A"
[8,]	"area-OJ-1-morning-A"	"area-VC-1-morning-A"
[9,]	"area-OJ-2-morning-A"	"area-VC-2-morning-A"
[10,]	"area-OJ-0.5-afternoon-B"	"area-VC-0.5-afternoon-B"
[11,]	"area-OJ-1-afternoon-B"	"area-VC-1-afternoon-B"
[12,]	"area-OJ-2-afternoon-B"	"area-VC-2-afternoon-B"
[13,]	"area-OJ-0.5-evening-B"	"area-VC-0.5-evening-B"
[14,]	"area-OJ-1-evening-B"	"area-VC-1-evening-B"
[15,]	"area-OJ-2-evening-B"	"area-VC-2-evening-B"
[16,]	"area-OJ-0.5-morning-B"	"area-VC-0.5-morning-B"
[17,]	"area-OJ-1-morning-B"	"area-VC-1-morning-B"
[18,]	"area-OJ-2-morning-B"	"area-VC-2-morning-B"

```
, , 2
```

	[,1]	[,2]
[1,]	"len-OJ-0.5-afternoon-A"	"len-VC-0.5-afternoon-A"
[2,]	"len-OJ-1-afternoon-A"	"len-VC-1-afternoon-A"
[3,]	"len-OJ-2-afternoon-A"	"len-VC-2-afternoon-A"
[4,]	"len-OJ-0.5-evening-A"	"len-VC-0.5-evening-A"
[5,]	"len-OJ-1-evening-A"	"len-VC-1-evening-A"
[6,]	"len-OJ-2-evening-A"	"len-VC-2-evening-A"
[7,]	"len-OJ-0.5-morning-A"	"len-VC-0.5-morning-A"
[8,]	"len-OJ-1-morning-A"	"len-VC-1-morning-A"
[9,]	"len-OJ-2-morning-A"	"len-VC-2-morning-A"
[10,]	"len-OJ-0.5-afternoon-B"	"len-VC-0.5-afternoon-B"
[11,]	"len-OJ-1-afternoon-B"	"len-VC-1-afternoon-B"
[12,]	"len-OJ-2-afternoon-B"	"len-VC-2-afternoon-B"
[13,]	"len-OJ-0.5-evening-B"	"len-VC-0.5-evening-B"
[14,]	"len-OJ-1-evening-B"	"len-VC-1-evening-B"
[15,]	"len-OJ-2-evening-B"	"len-VC-2-evening-B"
[16,]	"len-OJ-0.5-morning-B"	"len-VC-0.5-morning-B"

```
[17,]  "len-OJ-1-morning-B"        "len-VC-1-morning-B"
[18,]  "len-OJ-2-morning-B"        "len-VC-2-morning-B"


, , 3


[,1]                              [,2]
[1,]   "weight-OJ-0.5-afternoon-A" "weight-VC-0.5-afternoon-A"
[2,]   "weight-OJ-1-afternoon-A"   "weight-VC-1-afternoon-A"
[3,]   "weight-OJ-2-afternoon-A"   "weight-VC-2-afternoon-A"
[4,]   "weight-OJ-0.5-evening-A"   "weight-VC-0.5-evening-A"
[5,]   "weight-OJ-1-evening-A"     "weight-VC-1-evening-A"
[6,]   "weight-OJ-2-evening-A"     "weight-VC-2-evening-A"
[7,]   "weight-OJ-0.5-morning-A"   "weight-VC-0.5-morning-A"
[8,]   "weight-OJ-1-morning-A"     "weight-VC-1-morning-A"
[9,]   "weight-OJ-2-morning-A"     "weight-VC-2-morning-A"
[10,]  "weight-OJ-0.5-afternoon-B" "weight-VC-0.5-afternoon-B"
[11,]  "weight-OJ-1-afternoon-B"   "weight-VC-1-afternoon-B"
[12,]  "weight-OJ-2-afternoon-B"   "weight-VC-2-afternoon-B"
[13,]  "weight-OJ-0.5-evening-B"   "weight-VC-0.5-evening-B"
[14,]  "weight-OJ-1-evening-B"     "weight-VC-1-evening-B"
[15,]  "weight-OJ-2-evening-B"     "weight-VC-2-evening-B"
[16,]  "weight-OJ-0.5-morning-B"   "weight-VC-0.5-morning-B"
[17,]  "weight-OJ-1-morning-B"     "weight-VC-1-morning-B"
[18,]  "weight-OJ-2-morning-B"     "weight-VC-2-morning-B"
```

where the column names using identical supplement are arranged column-wisely for each data level ("area", "len" and "weight"), while the combinations of other unfixed layer (dose, supplementation time and nutrition company) are listed row by row using the identical sort.

This design overcomes the demerits under the two mentioned situations: multiple types of data and multi assignment for fixed factors. It eliminates the dependency between function's body and the data and variable types, through which a generic group operation on dataset can be implemented. Regardless of the numbers of data types and factorial variables, after the tensor is rearranged, most group operations on the dataset based on multiple fixed factors can be achieved in a customized function only using a double-nested loop structure:

```
GroupOps <- function(x, ...){
  Opsarr <- grparr(x, ...)
  dims <- dim(Opsarr)
  for (i in 1:dims[3]) {
    for (j in 1:dims[2]) {
      subset1 <- x[,Opsarr[,j,i]]
      "Your Operation on data type i, factorial combination j"
      ...
    }
  }
}
```

```
  . . .
}
```

4   Discussion and Conclusion

One of the advantages of using the VBTree style data frame is the lower memory usage, compared with the conventional style data frame in R. Take the dataset ToothGrowth as an example:

```
> lapply(list(ToothGrowth, TGvbt), object.size)
[[1]]
2696 bytes

[[2]]
2144 bytes
```

It is obvious that although both of them cover the same amount of information, the VBTree style data frame (TGvbt) takes up only 80% the size of ToothGrowth.

Another advantage of the VBTree's solution for group operations on dataset is that it was implemented by being manipulated on column names. This strategy eliminates the dependency between the scale of original dataset and the specific data processing operations so that it is unnecessary for the program to visit the original dataset frequently before the group processing. Most importantly, all data could be driven by the following three VBTree intrinsic data structures with the invariant sizes which are irrespective of the numbers of total rows (namely, the size of the dataset):

```
> dl <- chrvec2dl(colnames(TGvbt))
> vbt <- dl2vbt(dl)
> arr <- vbt2arr(vbt)
> lapply(list(dl, vbt, arr), object.size)
[[1]]
848 bytes

[[2]]
1456 bytes

[[3]]
1016 bytes
```

Thus, it is expected that the more rows the dataset has, the faster the algorithm will perform using VBTree strategies.

All the mentioned advantages of th VBTree package render it a novel, feasible and flexible method for group operations on datasets in R. Using VBTree, we aim to develope an easy-to-use optional solution for big data calculation, for not only industrial but also scientific communities.

5 Acknowledgement

This project was supported by the Chinese Government Graduate Student Overseas Study Program (2016), China Scholarship Council (CSC).

References

Becker, R. A., Chambers, J. M., & Wilks, A. R. (1988). The new s language. wadsworth & brooks. *Cole publication*.

Bliss, C. (1952). The statistics of bioassay, acad. press, new york. *Vitamin Methods*, *2*, 445–628.

Draper, N. R., & Smith, H. (2014). *Applied regression analysis* (Vol. 326). John Wiley & Sons.

On the Relationship between Factor Analysis and Principal Component Analysis in High-Dimensions[*]

Kentaro Hayashi[1] and Ke-Hai Yuan[2,3]

[1] University of Hawaii at Manoa, USA
hayashik@hawaii.edu
[2] University of Notre Dame, USA
[3] Nanjing University of Posts and Telecommunications, China
kyuan@nd.edu

Abstract. This article reviews the relationship between loadings from factor analysis (FA) and those from principal component analysis (PCA) when the number of variables p is large. While FA and PCA are substantively different methodologies, the two loading matrices are often close to each other. After defining how to measure the degree of closeness between the two loading matrices under rotational indeterminacy, the article reviews the conditions under which the two loading matrices agree with each other. Two well-known conditions for characterizing the two sets of loadings are given by Guttman (1956) and by Schneeweiss (1997), and they are further refined by Krijnen (2006). The relationship between these conditions is discussed, and results are provided showing that the two loading are closely related when p is large. Estimation methods are described to deal with conditions when sparsity does not hold in the covariance matrix and when p is not much greater than the sample size. Also, the problem of an increased bias in eigenvalues of the covariance matrix as p increases is also noted.

Keywords: Bias in eigenvalues · Canonical correlation · Matrix norm.

DOI: 10.35566/isdsa2019c9

1 Factor Analysis and Principal Component Analysis

Factor analysis (FA) (Anderson, 2003; Lawley & Maxwell, 1971) and principal component analysis (PCA) (Anderson, 2003; Jolliffe, 2002) are frequently used multivariate statistical methods for data reduction, especially in social and behavioral sciences. Oftentimes, PCA is used to approximate FA, and an important research question is under what condition PCA gives a good approximation to FA.

---

[*] This work was supported by Grant 31971029 from the Natural Science Foundation of China.

We start with defining the two methods. In FA, the p-dimensional mean-centered vector of the observed variables $\mathbf{y}$ is linearly related to an m-dimensional vector of latent factors $\mathbf{f}$ according to

$$\mathbf{y} = \mathbf{\Lambda f} + \varepsilon, \tag{1}$$

where $\mathbf{\Lambda}$ is a $p \times m$ matrix of factor loadings (with $p > m$), and ε is a p-dimensional vector of errors. Typically, for the orthogonal factor model in which factors are uncorrelated, three assumptions are imposed: (i) $\mathbf{f} \sim N_m(\mathbf{0}, \mathbf{I})$; (ii) $\varepsilon \sim N_p(\mathbf{0}, \mathbf{\Psi})$, where $\mathbf{\Psi}$ is a diagonal matrix with positive diagonal elements; (iii) $\mathrm{Cov}(\mathbf{f}, \varepsilon) = \mathbf{0}$. In words, factors and errors are normally distributed; errors corresponding to different observed variables are uncorrelated; and there are no correlations between factors and errors. Then, under these three assumptions, the covariance matrix of $\mathbf{y}$ is given by

$$\mathbf{\Sigma} = \mathbf{\Lambda\Lambda}' + \mathbf{\Psi}. \tag{2}$$

For PCA, we also assume that $\mathbf{y}$ has a covariance matrix $\mathbf{\Sigma}$. The first principal component (PC) is the linear combination of the elements of $\mathbf{y}$ that has the maximum variance subject to the condition that the l_2 norm of the vector of coefficients equals 1. The second PC is the linear combination of $\mathbf{y}$ that has the maximum variance subject to the condition that the l_2 norm of the vector of coefficients equals 1 and the constraint that it is uncorrelated with the first PC. Likewise, the ith PC is the linear combination of the elements of $\mathbf{y}$ that has the maximum variance subject to the condition that the l_2 norm of the vector of coefficients equals 1 and the constraints that it is uncorrelated with all the previous $(i-1)$ PCs. It is well known that the problem of finding such PCs reduces to the problem of finding the eigenvalues and standardized eigenvectors of $\mathbf{\Sigma}$. Let $\mathbf{\Lambda}^+$ be the $p \times m$ matrix whose columns are the standardized eigenvectors corresponding to the first m largest eigenvalues of $\mathbf{\Sigma}$; $\mathbf{\Omega}$ be the $m \times m$ diagonal matrix whose diagonal elements are the first m largest eigenvalues of $\mathbf{\Sigma}$. Then the first m PCs are obtained as

$$\mathbf{f}^* = \mathbf{\Lambda}^{+\prime}\mathbf{y}. \tag{3}$$

Clearly, the PCs are uncorrelated with $\mathrm{Cov}(\mathbf{f}^*) = \mathbf{\Lambda}^{+\prime}\mathbf{\Sigma}\mathbf{\Lambda}^+ = \mathbf{\Omega}$. When m is properly chosen, there exists

$$\mathbf{\Sigma} \approx \mathbf{\Lambda}^+\mathbf{\Omega}\mathbf{\Lambda}^{+\prime} = \mathbf{\Lambda}^*\mathbf{\Lambda}^{*\prime}, \tag{4}$$

where $\mathbf{\Lambda}^* = \mathbf{\Lambda}^+\mathbf{\Omega}^{1/2}$ is the $p \times m$ matrix of PCA loadings, with $\mathbf{\Omega}^{1/2}$ being the $m \times m$ diagonal matrix whose diagonal elements are the square root of those in $\mathbf{\Omega}$. Obviously, we can express (4) as

$$\mathbf{\Sigma} = \mathbf{\Lambda}^*\mathbf{\Lambda}^{*\prime} + \mathbf{\Psi}^*, \tag{5}$$

where $\mathbf{\Psi}^*$ is a function of the smallest $(p-m)$ eigenvalues and the corresponding standardized eigenvectors of $\mathbf{\Sigma}$. More specifically, $\mathbf{\Psi}^* = \mathbf{\Lambda}_1^+\mathbf{\Omega}_1\mathbf{\Lambda}_1^{+\prime}$, where $\mathbf{\Omega}_1$ is the $(p-m) \times (p-m)$ diagonal matrix whose elements are the smallest $(p-m)$ eigenvalues of $\mathbf{\Sigma}$ arranged in descending order, and $\mathbf{\Lambda}_1^+$ ($p \times (p-m)$) is the corresponding standardized eigenvectors.

2 Differences between Factor Analysis and Principal Component Analysis

Based on the above definitions, FA and PCA are substantially different statistical methods. In particular, their major differences include:

(a) First of all, from equations (1) and (3), we instantly notice that the role of observed variables and factors/PCs are reversed. In FA, the observed variables are dependent variables and the factors are explanatory variables. On the contrary, in PCA, the observed variables are explanatory variables and PCs are dependent variables. This is by far the largest difference between the two methods.

(b) FA is a statistical model, while PCA is simply the eigenvalue-eigenvector decomposition of Σ. Because FA is a statistical model, there is a corresponding null hypothesis that can be tested. Also, factors and errors in FA are assumed uncorrelated. On the other hand, more variance is explained as the number of PCs increases, and there are no distributional assumptions in PCA.

(c) Though both FA and PCA are used for data reduction, the primary focus of FA is to model the covariance structure, whereas PCA focuses more on data reduction.

(d) The PCs are uniquely defined as a result of the eigenvalue-eigenvector decomposition of Σ, while factors cannot be defined in a unique way. There are different factor score estimators. The two most famous ones are the Bartlett estimator (Bartlett, 1937) and the regression estimator, however, there are other factor score estimators (Krijnen, Wansbeek, & ten Berge, 1996; ten Berge, Krijnen, Wansbeek, & Shapiro, 1999). Actually, factor score estimation has been an ongoing research topic in FA.

(e) FA parameters must be estimated iteratively with some algorithm, while sample PCA loadings can be found by simply applying the eigenvalue-eigenvector decomposition of the sample covariance matrix $\mathbf{S}$.

(f) Because the matrix of unique variances is positive definite, we encounter the problem of improper solution (also called the Heywood cases) (van Driel, 1978; Kano, 1998) in FA, while in PCA, there is no such complexity.

(g) The error variance matrix (i.e., the matrix of unique variances $\mathbf{\Psi}$) is diagonal in FA, while in PCA, the "error" variance matrix $\mathbf{\Psi}^*$ constructed from the smallest $(p-m)$ eigenvalues and corresponding eigenvectors is in general not a diagonal matrix.

3 Measures of Closeness between Factor Analysis and Principal Component Analysis

In spite of the substantial differences described above, it is well-known that FA and PCA often yield approximately the same results, especially their estimated loading matrices $\hat{\mathbf{\Lambda}}$ and $\hat{\mathbf{\Lambda}}^*$, respectively (e.g., Velicer & Jackson, 1990). Before discussing conditions under which the two matrices are close to each other, we must discuss how to measure the degree of closeness between FA and PCA loadings. This is because loading matrices are defined only when the indeterminacy is removed. The indeterminacy is due to (i) an orthogonal rotation, (ii) column alignments, and (iii) a sign change for each column. Therefore, it is necessary to take into account such indeterminacy in measuring the degree of closeness between FA and PCA loadings. Specifically, because the covariance structure under the FA model is given as in equation (2), the rotational indeterminacy makes it difficult to measure the closeness between FA and PCA loadings:

$$\mathbf{\Sigma} = \mathbf{\Lambda}\mathbf{T}\mathbf{T}'\mathbf{\Lambda}' + \mathbf{\Psi} = \mathbf{\Lambda}\mathbf{T}(\mathbf{\Lambda}\mathbf{T})' + \mathbf{\Psi} = \tilde{\mathbf{\Lambda}}\tilde{\mathbf{\Lambda}}' + \mathbf{\Psi}, \tag{6}$$

where $\tilde{\mathbf{\Lambda}} = \mathbf{\Lambda}\mathbf{T}$ and $\mathbf{T}$ is any $m \times m$ orthogonal rotation matrix. One solution might be to compare $\tilde{\mathbf{\Lambda}} = \mathbf{\Lambda}\mathbf{T}$ and $\mathbf{\Lambda}^*$, and to take the minimized squared matrix norm (Ichikawa & Konishi, 1995) across all $\mathbf{T}$ (i.e., $\min_{\mathbf{T}:\mathbf{T}'\mathbf{T}=\mathbf{I}_m}\{\mathrm{tr}[(\mathbf{\Lambda}\mathbf{T} - \mathbf{\Lambda}^*)'(\mathbf{\Lambda}\mathbf{T}-\mathbf{\Lambda}^*)]\}$) as the degree of closeness, where $\mathrm{tr}(\mathbf{A})$ is the trace (sum of the diagonal elements) of a square matrix $\mathbf{A}$. Though in the FA literature, orthogonal rotations were developed as a technique to find a simple structure (Thurstone, 1947), column alignments and a sign change of each column of loadings are special cases of orthogonal rotations. Suppose $m = 2$, then a post-multiplication of a loading matrix by

$$\mathbf{T} = \begin{pmatrix} 0 & 1 \\ 1 & 0 \end{pmatrix} \quad \text{and} \quad \mathbf{T} = \begin{pmatrix} 1 & 0 \\ 0 & -1 \end{pmatrix}$$

takes care of the column exchange and the sign change of the second column, respectively. Obviously, both matrices are orthogonal matrices (See, e.g., Section 8.4 of Harvelle, 1997).

Suppose that the issue with rotational indeterminacy between $\mathbf{\Lambda}$ and $\mathbf{\Lambda}^*$ or $\hat{\mathbf{\Lambda}}$ and $\hat{\mathbf{\Lambda}}^*$ have been resolved. Then the squared matrix (Forbenius) norm (See e.g., p. 291 of Horn & Johnson, 1985; p. 165 of Schott, 2005) of the difference between matrices $\mathbf{\Lambda}$ $(p \times m)$ and $\mathbf{\Lambda}^*$ $(p \times m)$ is given by

$$||\mathbf{\Lambda} - \mathbf{\Lambda}^*||^2 = \mathrm{tr}\{(\mathbf{\Lambda} - \mathbf{\Lambda}^*)'(\mathbf{\Lambda} - \mathbf{\Lambda}^*)\}. \tag{7}$$

A way to avoid rotational indeterminacy is to employ the average squared canonical correlations between the two loading matrices $\mathbf{\Lambda}$ and $\mathbf{\Lambda}^*$ instead of an ordinary matrix norm. The squared canonical correlations are a measure of closeness between $\mathbf{\Lambda}$ and $\mathbf{\Lambda}^*$ (Schneeweiss & Mathes, 1995; Schneeweiss, 1997). They are given by the eigenvalues of $(\mathbf{\Lambda}'\mathbf{\Lambda})^{-1}(\mathbf{\Lambda}'\mathbf{\Lambda}^*)(\mathbf{\Lambda}^{*'}\mathbf{\Lambda}^*)^{-1}(\mathbf{\Lambda}^{*'}\mathbf{\Lambda})$, and are known to be invariant with respect to orthogonal rotations. Thus, the average squared canonical correlation between matrices $\mathbf{\Lambda}$ and $\mathbf{\Lambda}^*$ is given by

$$\rho^2(\mathbf{\Lambda}, \mathbf{\Lambda}^*) = \frac{\mathrm{tr}\{(\mathbf{\Lambda}'\mathbf{\Lambda})^{-1}(\mathbf{\Lambda}'\mathbf{\Lambda}^*)(\mathbf{\Lambda}^{*'}\mathbf{\Lambda}^*)^{-1}(\mathbf{\Lambda}^{*'}\mathbf{\Lambda})\}}{m}. \tag{8}$$

For a one-factor model, $\rho^2(\mathbf{\Lambda}, \mathbf{\Lambda}^*)$ reduces to:

$$\begin{aligned} \rho^2(\boldsymbol{\lambda}, \boldsymbol{\lambda}^*) &= (\boldsymbol{\lambda}'\boldsymbol{\lambda})^{-1}(\boldsymbol{\lambda}'\boldsymbol{\lambda}^*)(\boldsymbol{\lambda}^{*'}\boldsymbol{\lambda}^*)^{-1}(\boldsymbol{\lambda}^{*'}\boldsymbol{\lambda}) \\ &= \frac{(\boldsymbol{\lambda}'\boldsymbol{\lambda}^*)(\boldsymbol{\lambda}^{*'}\boldsymbol{\lambda})}{(\boldsymbol{\lambda}'\boldsymbol{\lambda})(\boldsymbol{\lambda}^{*'}\boldsymbol{\lambda}^*)} = \frac{(\boldsymbol{\lambda}'\boldsymbol{\lambda}^*)^2}{||\boldsymbol{\lambda}||_2^2||\boldsymbol{\lambda}^*||_2^2} = \{\mathrm{Corr}(\boldsymbol{\lambda}, \boldsymbol{\lambda}^*)\}^2. \end{aligned} \tag{9}$$

That is, for the one-factor model, the squared canonical correlation reduces to the squared correlation between $\boldsymbol{\lambda}$ and $\boldsymbol{\lambda}^*$. According to the Cauchy-Schwarz inequality, for a one-factor model, a perfect (limiting) squared correlation

$$\rho^2(\boldsymbol{\lambda}, \boldsymbol{\lambda}^*) = \frac{(\boldsymbol{\lambda}'\boldsymbol{\lambda}^*)^2}{||\boldsymbol{\lambda}||_2^2||\boldsymbol{\lambda}^*||_2^2} \to 1$$

holds if and only if $\boldsymbol{\lambda}^* - c\boldsymbol{\lambda} \to \mathbf{0}$ with some non-zero constant c.

A lower bound of the average squared canonical correlation can be given in terms of the squared matrix norm under a common assumption of $(m/p) \to 0$ as

$$\rho^2(\boldsymbol{\Lambda}, \boldsymbol{\Lambda}^*) \geq 1 - \frac{C}{m}||\boldsymbol{\Lambda} - \boldsymbol{\Lambda}^*||^2 \tag{10}$$

for some constant $C > 0$ (Hayashi, Yuan, & Liang, 2017). Using Monte Carlo simulation, Liang, Hayashi, and Yuan (2015) showed that the value of the Fisher-z transformed average canonical correlation ($\hat{z} = (1/2)\log\{[1 + \rho(\hat{\boldsymbol{\Lambda}}, \hat{\boldsymbol{\Lambda}}^*)]/[1 - \rho(\hat{\boldsymbol{\Lambda}}, \hat{\boldsymbol{\Lambda}}^*)]\}$) and the logarithm of p (i.e., $\log(p)$) are almost perfectly linearly related. Also, Liang, Hayashi, and Yuan (2016) found in their simulation study that, under high-dimensional conditions, the closeness between the estimated FA and the estimated PCA loadings is achieved faster than that between the estimated and the population FA loadings and between the estimated PCA and the population FA loadings.

4   High Dimensions or Semi-High Dimensions

With the advancement of computing technology and different data collection toolboxes, high-dimensional data (with a large p) arise in many disciplines (See, e.g., Tibshirani, 1996; Hastie, Tibshirani, & Friedman, 2009; Buehlmann & van de Geer, 2011). Consequently, the needs for improving the statistical methodology for analyzing such data are increasing. Large p is also common in the traditional research in social sciences, in which we often collect data from multiple questionnaires. As a result, whenever conducting item analysis with multiple questionnaires combined, we have to face the issue of analyzing high-dimensional data.

Under high-dimensionality, a common approach is to assume sparsity in the covariance matrix (i.e., assuming that many off-diagonal elements are either zero or near zero) and to apply some regularized methods. See, e.g., Pourahmadi (2013), and Engel, Buydens and Blanchet (2017) for an overview on estimating high-dimensional covariance matrix. Under high-dimensional setting and when both p and n approach infinity together with $(\sqrt{p}/n) \to 0$, Bai and Li (2012) showed that the loading estimates by FA and PCA converge to the same asymptotic normal distributions.

However, in most applications in the social sciences, the covariance matrices are not necessarily sparse. Consequently, the assumption of a sparse covariance matrix may not hold. Also, in the social sciences, even when the number of variables p is greater than the sample size n, they are still comparable in sizes. We rarely encounter situations with the dimension far exceeding the sample size (i.e., "$p >> n$"). We call such situations (i.e., (i) the covariance matrix is not sparse, and (ii) p and n are comparable in size though p can be larger than n) "semi-high" dimensional settings to distinguish them from the high dimensional settings encountered in e.g., statistical learning in recent years. It is consistent with the conditions examined by statisticians studying multivariate analysis in high dimensions, e.g., $(p/n) \to c \in (0, 1)$ as $n \to \infty$ and $p \to \infty$ (e.g., Ulyanov, Wakaki, & Fujikoshi, 2006; Wakaki, Fujikoshi, & Ulyanov 2014).

In short, although some authors called data with $p > n$ as high-dimensional (See, e.g., Hastie, Tibshirani, & Friedman, 2009, Chapter 18; Pourahmadi, 2013), we do not require this assumption to accommodate typical social science data. Also, we do not consider it a requirement for the covariance matrix to be sparse.

5 Conditions for Closeness between Factor Analysis and Principal Component Analysis

To discuss conditions under which the two matrices (i.e., Λ and Λ^*) are close to each other, we must first discuss some common assumptions (See, e.g., Krijnen, 2006). This is because if all the unique variances vanish, then it is trivially true that FA and PCA loadings agree with each other. However, this violates the assumption in FA that the matrix of unique variances Ψ is positive definite (i.e., all the elements of unique variances are positive). As noted earlier, the violation of this assumption at the level of the estimates is called either an improper solution or the Heywood case.

We write the ith diagonal element of Σ as σ_{ii}, and denote the supremum of σ_{ii} $(i = 1, \ldots, p)$ as $\sigma_{\sup}$ (i.e., $\sigma_{\sup} = \sup_{i \geq 1} \sigma_{ii}$). Likewise, we write the ith diagonal element of Ψ as ψ_{ii}, and denote the supremum and the infimum of ψ_{ii} as $\psi_{\sup} = \sup_{i \geq 1} \psi_{ii}$ and $\psi_{\inf} = \inf_{i \geq 1} \psi_{ii}$, respectively. Krijnen (2006) assumed that the supremum of the diagonal elements of Σ is finite (i.e., $\sigma_{\sup} < \infty$) and that the infimum of the unique variances is bounded away from zero $(0 < \psi_{\inf})$. Then obviously, $0 < \psi_{\inf} \leq \psi_{ii} \leq \psi_{\sup} \leq \sigma_{\sup} < \infty$. The direct consequence of this assumption is that not only the diagonal elements of Ψ are bounded above and also bounded away from zero, but the elements of the inverse matrix Ψ^{-1} are also bounded above and bounded away from zero. That is, $0 < \psi_{\sup}^{-1} \leq \psi_{ii}^{-1} \leq \psi_{\inf}^{-1} < \infty$. Furthermore, we denote the infimum of σ_{ii} as $\sigma_{\inf}$ (i.e., $\sigma_{\inf} = \inf_{i \geq 1} \sigma_{ii}$). Because $0 < \psi_{\inf} \leq \sigma_{\inf}$, $\sigma_{\inf}$ is also bounded away from zero. In addition, we denote $d_{\min}$ as the minimum eigenvalue(s) of $\Lambda'\Lambda$. (Note: If we let $m \to \infty$ as $p \to \infty$, we must instead define the infimum of the eigenvalues of $\Lambda'\Lambda$.)

With these notations, we next discuss conditions under which $\hat{\Lambda}$ and $\hat{\Lambda}^*$ are close to each other. At the population level, two such conditions identified by Guttman (1956) and Schneeweiss (1997) are most well-known.

(a) Guttman (1956)

As in equation (2), the covariance structure under the FA model is given as $\mathbf{\Sigma} = \mathbf{\Lambda}\mathbf{\Lambda}' + \mathbf{\Psi}$, where $\mathbf{\Psi}$ is a diagonal unique variance matrix. Let $\mathbf{\Sigma}^{-1} = (\sigma^{ij})$ and $\mathbf{\Psi} = \text{diag}(\psi_{11}, \psi_{22}, \ldots, \psi_{pp})$. Guttman (1956; See also Theorem 1 of Krijnen, 2006) showed that if $(m/p) \to 0$ as $p \to \infty$, then $\psi_{jj}\sigma^{jj} \to 1$ for almost all j. Here, "for almost all j" means $\lim_{p \to \infty} \#\{j : \psi_{jj}\sigma^{jj} < 1\}/p = 0$. That is, the number of j that satisfies $\psi_{jj}\sigma^{jj} < 1$ is ignorable as p goes to infinity. Note that the Guttman condition is not explicitly on the closeness between FA and PCA loadings. However, in fact, the condition is closely related to the closeness between the two loading matrices (Hayashi & Bentler, 2000; Krijnen, 2006).

(b) Bentler and Kano (1990)

For the one-factor model with $\mathbf{\Lambda} = \boldsymbol{\lambda}$ and $\mathbf{\Lambda}^* = \boldsymbol{\lambda}^*$, under the conditions that $\boldsymbol{\lambda}'\boldsymbol{\lambda} \to \infty$ and there exists an upper bound for unique variances as $p \to \infty$, Bentler and Kano (1990) proved that $\boldsymbol{\lambda}^*$ converges to $\boldsymbol{\lambda}$. To the best of our knowledge, the technical work of Bentler and Kano (1990) is the first to give an explicit condition for the closeness between FA and PCA. We consider Schneeweiss and Mathes (1995) and Schneeseiss (1997) as an extension to Bentler and Kano (1990).

(c) Schneeweiss and Mathes (1995) and Schneeweiss (1997)

Schneeweiss and Mathes (1995) showed that $\rho^2(\mathbf{\Lambda}, \mathbf{\Lambda}^*) \to 1$ if $d_{\min}^{-1}\psi_{\sup} \to 0$. Schneeweiss (1997) further gave a weaker condition for $\rho^2(\mathbf{\Lambda}, \mathbf{\Lambda}^*) \to 1$: $d_{\min}^{-1}\delta \to 0$ where $\delta = \psi_{\sup} - \psi_{\inf}$ is the difference between the supremum and the infimum of the diagonal elements of $\mathbf{\Psi}$. Here, note that the Guttman condition is expressed by only p and m, and the role of FA loadings is not mentioned. On the other hand, the Schneeweiss-Mathes and Schneeweiss conditions are expressed in terms of the eigenvalue(s) of FA loadings and unique variance(s), and the roles of p and m are not mentioned. Yet, it is known that these two sets of conditions are closely related.

With the assumption that $0 < \psi_{\sup}^{-1} \le \psi_{\inf}^{-1} < \infty$, the closeness condition between the loading matrix from FA and that from PCA by Schneeweiss and Mathes (1995) can be stated as $ev_m(\mathbf{\Lambda}'\mathbf{\Psi}^{-1}\mathbf{\Lambda}) \to \infty$, where $ev_k(\mathbf{A})$ is the kth largest eigenvalue of a square matrix $\mathbf{A}$. Obviously, $ev_m(\mathbf{\Lambda}'\mathbf{\Psi}^{-1}\mathbf{\Lambda})$ is the smallest eigenvalue of $\mathbf{\Lambda}'\mathbf{\Psi}^{-1}\mathbf{\Lambda}$.

Related to the Schneeweiss condition, Bentler (1976) parameterized the correlation structure of the factor model as $\mathbf{\Psi}^{-1/2}\mathbf{\Sigma}\mathbf{\Psi}^{-1/2} = \mathbf{\Psi}^{-1/2}(\mathbf{\Lambda}\mathbf{\Lambda}')\mathbf{\Psi}^{-1/2} + \mathbf{I}_p$ and showed that, under this parameterization, a necessary condition for $ev_m(\mathbf{\Lambda}'\mathbf{\Psi}^{-1}\mathbf{\Lambda}) = ev_m(\mathbf{\Psi}^{-1/2}\mathbf{\Lambda}\mathbf{\Lambda}'\mathbf{\Psi}^{-1/2}) \to \infty$ is that, as p increases, the sum of squared loadings on each factor has to go to infinity ($\boldsymbol{\lambda}_k'\boldsymbol{\lambda}_k \to \infty$, $k = 1, \ldots, m$, as $p \to \infty$).

(d) Relationship between Guttman and Schneeweiss Conditions

The relationship between the Guttman and Schneeweiss conditions is summarized in Table 1 (Hayashi, Yuan, & Jiang, 2019). In particular, Schneeweiss condition ($ev_m(\mathbf{\Lambda}'\mathbf{\Psi}^{-1}\mathbf{\Lambda}) \to \infty$) is sufficient for Guttman condition (($m/p) \to 0$ as $p \to \infty$) (Krijnen, 2006, Theorem 3). The converse (($m/p) \to 0$ as $p \to \infty$ implies $ev_m(\mathbf{\Lambda}'\mathbf{\Psi}^{-1}\mathbf{\Lambda}) \to \infty$) also holds with an extra condition that the points of measure zero can be ignored so that "for *almost* all $j = 1, \ldots, p$" is replaced by "for all $j = 1, \ldots, p$", as follows.

Table 1. Relationships among Conditions and Results (Hayashi, Yuan, & Jiang, 2019).

	Condition	Result	Source
1	$(m/p) \to 0$ as $p \to \infty$	$\psi_{jj}\sigma^{jj} \to 1$ for almost all j	Guttman (1956) Krijnen (2006, Thm 1)
2	$\psi_{jj}\sigma^{jj} \to 1$ for almost all j $ev_m(\Lambda'\Psi^{-1}\Lambda) > c > 0$	$(m/p) \to 0$ as $p \to \infty$	Krijnen (2006, Thm 2)
3	$ev_m(\Lambda'\Psi^{-1}\Lambda) \to \infty$	$(m/p) \to 0$ as $p \to \infty$	Krijnen (2006, Thm 3) Hayashi & Bentler (2000, Obs 8b)
4	$ev_m(\Lambda'\Psi^{-1}\Lambda) \to \infty$	$\psi_{jj}\sigma^{jj} \to 1$ for all j	Krijnen (2006, Thm 4) Hayashi & Bentler (2000, after Obs 7)
5	$\psi_{jj}\sigma^{jj} \to 1$ for all j	$ev_m(\Lambda'\Psi^{-1}\Lambda) \to \infty$	Krijnen (2006, Thm 4)
6	$ev_j(\Lambda'\Psi^{-1}\Lambda) \to \infty$ $ev_k(\Lambda'\Psi^{-1}\Lambda) < C < \infty$ $j = 1, \ldots, r$ $k = r+1, \ldots, m$ $p \to \infty$; m fixed	$\psi_{jj}\sigma^{jj} \to 1$ for almost all j	Krijnen (2006, Thm 5)
7	$ev_m(\Lambda'\Psi^{-1}\Lambda) \to \infty$	$\Psi^{-1} - \Sigma^{-1} \to 0$	Hayashi, Yuan, & Jiang (20019)

Notes: (i) 2 is a partial converse of 1; (ii) 5 is the converse of 4; (iii) 7 is a matrix version of 4.

First, the condition of $(m/p) \to 0$ as $p \to \infty$ is sufficient for $\psi_{jj}\sigma^{jj} \to 1$ for almost all j (Guttman, 1956; Krijnen, 2006, Theorem 1). Also, $\psi_{jj}\sigma^{jj} \to 1$ for all j implies $ev_m(\Lambda'\Psi^{-1}\Lambda) \to \infty$ (Krijnen, 2006, Theorem 4). Here, "$\psi_{jj}\sigma^{jj} \to 1$ for all j" is slightly stronger than "$\psi_{jj}\sigma^{jj} \to 1$ for almost all j." However, in practice, it seems reasonable to assume that the number of loadings on every factor increases with p proportionally, as stated in Bentler (1976). Then the condition of $(m/p) \to 0$ as $p \to \infty$ becomes equivalent to $ev_m(\Lambda'\Psi^{-1}\Lambda) \to \infty$. That is, the Guttman and Schneeweiss conditions become interchangeable. See Table 1 for the summary of the relationship among different conditions.

6   Estimation

If we cannot assume sparsity of the covariance matrix, the most promising existing approach that can be used under semi-high dimensional settings seems to be the ridge-type estimation (e.g., Yuan & Chan, 2008; 2016). For example, Yuan and Chan (2008) employed a ridge maximum likelihood (ML) approach to estimate the parameters of structural equation models (SEM) which are a structured version of FA. They added a ridge constant $k > 0$ to the diagonals of $\mathbf{S}$, and used $\mathbf{S} + k\mathbf{I}_p$ in place of $\mathbf{S}$ in order to stably estimate the parameters in a FA-type model. If the dimension p exceeds the sample size n, the sample covariance matrix $\mathbf{S}$ is no longer positive definite and the inverse $\mathbf{S}^{-1}$ no longer exists. Then we cannot estimate the unique variances $\mathbf{\Psi}$ nor the inverse $\mathbf{\Psi}^{-1}$ under the FA model with either the generalized least squares (GLS) or the ML method. However, we can still employ the ridge ML and ridge GLS methods.

Hayashi, Yuan, and Jiang (2019, in press) suggested estimating the parameters of FA by using the unweighted least squares (ULS) method or performing the FA with equal unique variances (e.g., Hayashi & Bentler, 2000). Both the ULS and the FA with equal unique variances can be used without regularization. The ULS method minimizes the fit function $F_{ULS}(\mathbf{S}, \mathbf{\Sigma}) = \mathrm{tr}\{(\mathbf{S} - \mathbf{\Sigma})^2\}$ which does not involve either $\mathbf{S}^{-1}$ or the inverse of the estimated model-reproduced covariance matrix $\hat{\mathbf{\Sigma}}^{-1}$. Estimation with ULS is simpler than that with the GLS or the ML method.

The FA model with equal unique variances (with standardized variables) approximates the correlation matrix as:

$$\mathbf{\Sigma} \approx \mathbf{\Lambda}^*\mathbf{\Lambda}^{*\prime} + k\mathbf{I}_p, \tag{11}$$

where k is a positive constant and its ML estimate is given by the average of the $(p-m)$ smallest eigenvalues of $\mathbf{S}$ (i.e., $\hat{k} = (p-m)^{-1}\sum_{j=m+1}^{p} ev_j(\mathbf{S})$). Here, note that the eigenvectors of $\mathbf{\Sigma} - k\mathbf{I}_p$ are the same as the eigenvectors of $\mathbf{\Sigma}$, and the eigenvalues of $\mathbf{\Sigma} - k\mathbf{I}_p$ are simply those of $\mathbf{\Sigma}$ minus k. Thus, the FA model with equal unique variances can be considered as a variant of PCA. In fact, this model is also called the probabilistic PCA in statistics (Tipping & Bishop, 1999). Also, as was discussed, as p increases (with $(m/p) \to 0$), the loading matrices from the FA and the PCA have the same limiting values, up to a rotational indeterminacy. Thus, the FA model with equal unique variances seems a viable approach under the semi-high dimensional settings with a large p.

Through Monte Carlo simulation, Hayashi, Yuan, and Jiang (in press) showed that even when the sample size is smaller than the number of variables, the parameters of the FA model were still able to be estimated relatively accurately using the FA with equal unique variances, ULS, and ridge ML with small ridge constants. On average, the performances of these methods were nearly equal. When the values of the ridge constants were small, the ridge ML method worked fine, even though the selected values of the ridge constant were not optimized.

7   Should We Use Principal Component Analysis?

If FA and PCA loadings agree with each other in high dimensions, then we are inclined to use PCA rather than FA. This is because the estimation is simpler in PCA than in FA. Existing simulation results (Hayashi, Yuan, & Jiang, in press) indicate that FA with equal unique variances performed excellently. As we have pointed out, the FA with equal unique variances is a variant of PCA, also called the probabilistic PCA (Tipping & Bishop, 1999).

However, there might be a problem because the bias of the eigenvalues of Σ also increases as p increases (Hayashi, Yuan, & Liang, 2018). Suppose that $\mathbf{y}$ has a multivariate normal distribution with mean vector $\boldsymbol{\mu}$ and covariance matrix Σ, that is, $\mathbf{y} \sim N_p(\boldsymbol{\mu}, \Sigma)$. For a fixed value of p, Lawley (1956) showed that if the eigenvalues of Σ are distinct, that is, if $\omega_1 > \omega_2 > \ldots > \omega_p > 0$, then the mean of the ith largest eigenvalue l_i of the sample covariance matrix $\mathbf{S}$ can be expanded as

$$E(l_i) = \omega_i + \frac{\omega_i}{n} \sum_{j=1, j \neq i}^{p} \frac{\omega_j}{\omega_i - \omega_j} + O(n^{-2}). \tag{12}$$

By examining the bias term, Hayashi, Yuan, and Liang (2018) found that the order of the bias terms as a whole is $O(p/n)$. Therefore, we can write equation (12) as:

$$E(l_i) = \omega_i + O(p/n), \tag{13}$$

where $\omega_i = O(p)$ if $i = 1, \ldots, m$, and $\omega_i = O(1)$ if $i = m + 1, \ldots, p$.

Their results imply that (i) when p is ignorable relative to n, the order of the bias of the eigenvalues of the sample covariance matrix is $(1/n)$; (ii) when p is not ignorable relative to n, the order of the bias of the sample eigenvalues is (p/n); (iii) The sample eigenvalues are asymptotically unbiased if (p/n) goes to zero. It is obvious that the result in case (ii) is more general than that in case (i), and the sample eigenvalues are always asymptotically unbiased when $n \to \infty$ while holding p constant.

Here, it is important to note that problems might arise if $O(p/n)$ is not ignorable, in which case the assumption of $(p/n) \to 0$ does not hold. This indicates that the estimated eigenvalues are consistent if and only if $(p/n) \to 0$ (See also, e.g., the consistency result in Theorem 1 of Johnstone & Lu, 2009 on this point). Some remedies for the bias of eigenvalues of $\mathbf{S}$ have been proposed (See e.g., Arruda & Bentler, 2017 and references therein).

8   Summary

In this article, we reviewed the relationship between loadings from FA and PCA when the dimension p is large. We pointed out potential substantial differences between FA and PCA, although closeness between the loading matrices from FA and those from PCA is often observed. Our main focus was to discuss the conditions under which the two loading matrices agree with each other. We discussed the relationships among different conditions (especially between the

Guttman condition and the Schneenweiss condition), as well as the corresponding results when certain conditions hold under high dimensions (i.e., when p is large). We also discussed the estimation methods when sparsity does not hold in the covariance matrix and when p does not exceed n too much, and pointed out an increased bias in eigenvalues of the sample covariance matrix as p increases.

References

Anderson, T. W. (2003). *An introduction to multivariate statistical analysis* (3rd ed.). New York: Wiley.

Arruda, E. H. & Bentler, P. M. (2017). A regularized GLS for structural equation modeling. *Structural Equation Modeling, 24*, 657–665. doi: https://doi.org/10.1080/10705511.2017.1318392

Bai, J., & Li, K. (2012). Statistical analysis of factor models of high dimension. *Annals of Statistics, 40*, 436–465. doi: https://doi.org/10.1214/11-AOS966

Bartlett, M. S. (1937). The statistical conception of mental factors. *British Journal of Psychology, 28*, 97–104. doi: https://doi.org/10.1111/j.2044-8295.1937.tb00863.x

Bentler, P. M. (1976). Multistructure statistical model applied to factor analysis. *Multivariate Behavioral Research, 11*, 3–15. doi: https://doi.org/10.1207/s15327906mbr1101_1

Bentler, P. M., & Kano, Y. (1990). On the equivalence of factors and components. *Multivariate Behavioral Research, 25*, 67–74. doi: https://doi.org/10.1207/s15327906mbr2501_8

Buehlmann, P., & van de Geer, S. (2011). *Statistics for high-dimensional data: Method, theory, and applications.* Heidelberg: Springer. doi: https://doi.org/10.1007/978-3-642-20192-9

Engel, J., Buydens, L., & Blanchet, L. (2017). An overview of large-dimensional covariance and precision matrix estimator with applications in chemometrics. *Journal of Chemometrics, 31*, article e2880. doi: https://doi.org/10.1002/cem.2880

Guttman, L. (1956). "Best possible" systematic estimates of communalities. *Psychometrika, 21*, 273–286. doi: https://doi.org/10.1007/BF02289137

Harville, D. A. (1997). *Matrix algebra from a statistician's perspective.* New York: Springer. doi: https://doi.org/10.1080/00401706.1998.10485214

Hastie, T., Tibshirani, R., & Friedman, J. (2009). *The elements of statistical learning* (2nd ed.). New York: Springer. doi: https://doi.org/10.1007/978-0-387-84858-7_10

Hayashi, K., & Bentler, P. M. (2000). On the relations among regular, equal unique variances, and image factor analysis models. *Psychometrika, 65*, 59–72. doi: https://doi.org/10.1007/BF02294186

Hayashi, K., Yuan, K.-H., & Jiang, G. (2019). On extended Guttman condition in high dimensional factor analysis. In M. Wiberg, S. Culpepper, R. Janssen, J. Gonzalez, & D. Molenaar (Eds.), *Quantitative psychology: The 83rd annual meeting of the psychometric soci-*

ety,   New York City, 2018 (pp. 221–228). New York: Springer.   doi: https://link.springer.com/chapter/10.1007/978-3-030-01310-3_20

Hayashi, K., Yuan, K.-H., & Jiang, G. (in press).   On the precision matrix in semi-high dimensional settings. In M. Wiberg, J. Gonzalez, U. Bockenholt, & J.-S. Kim (Eds.), *Quantitative psychology: The 84th annual meeting of the psychometric society*, Santiago, Chile, 2019. New York: Springer.

Hayashi, K., Yuan, K.-H., & Liang, L. (2017).   On the relationship between squared canonical correlation and matrix norm. In L. A. van der Ark, M. Wiberg, S. Culpepper, J. A. Douglas, & W-C. Wang (Eds.), *Quantitative psychology: The 81st annual meeting of the psychometrics society*,   Asheville, NC, 2016 (pp. 141–150). New York: Springer.   doi: https://doi.org/10.1007/978-3-319-56294-0_13

Hayashi, K., Yuan, K.-H., & Liang, L. (2018).   On the bias in eigenvalues of sample covariance matrix. In M. Wiberg, S. Culpepper, R. Janssen, J. Gonzalez, & D. Molenaar (Eds.), *Quantitative psychology: The 82nd annual meeting of the psychometric society*, Zurich, Switzerland, 2017 (pp. 221–233). New York: Springer.   doi: https://doi.org/10.1007/978-3-319-77249-3_19

Horn, R. A., & Johnson, C. R. (1985). *Matrix analysis*. Cambridge: Cambridge University Press. doi: https://doi.org/10.1017/CBO9780511810817

Ichikawa, M. & Konishi, S. (1995).   Applications of the bootstrap methods in factor analysis. *Psychometrika, 60*, 77–93.   doi: https://doi.org/10.1007/BF02294430

Jolliffe, I. T. (2002). *Principal component analysis* (2nd ed.). New York: Springer. doi: https://doi.org/10.1002/0470013192.bsa501

Johnstone, I. M., & Lu, A. Y. (2009). On consistency and sparsity for principal components analysis in high dimensions. *Journal of the American Statistical Association, 104*, 682–693. doi: https://doi.org/10.1198/jasa.2009.0121

Kano Y. (1998). Improper solutions in exploratory factor analysis: Causes and treatments. In Rizzi A., Vichi M., Bock H. H. (Eds.), *Advances in data science and classification: Studies in classification, data analysis, and knowledge organization*. New York: Springer. doi: https://doi.org/10.1007/978-3-642-72253-0_51

Krijnen, W. P. (2006). Convergence of estimates of unique variances in factor analysis, based on the inverse sample covariance matrix. *Psychometrika, 71*, 193–199. doi: https://doi.org/10.1007/s11336-000-1142-9

Krijnen, W. P., Wansbeek, T. J., & ten Berge, J. M. F. (1996). Best linear predictors for factor scores. *Communications in Statistics: Theory and Methods, 25*, 3013–3025. doi: https://doi.org/10.1080/03610929608831883

Lawley, D. N. (1956). Test of significance for the latent roots of covariance and correlation matrices. *Biometrika, 43*, 128–136.   doi: https://doi.org/10.2307/2333586

Lawley, D. N., & Maxwell, A. E. (1971).   *Factor analysis as a statistical method* (2nd ed.).   New York: American Elsevier.   doi: https://doi.org/10.2307/2986915

Liang, L., Hayashi, K., & Yuan, K.-H. (2015). On closeness between factor analysis and principal component analysis under high-dimensional conditions. In L. A. van der Ark, D.M. Bolt, W-C. Wang, J. A. Douglas, & S-M. Chow (Eds.), *Quantitative psychology research: The 79th Annual Meeting of the Psychometric Society*, Madison, Wisconsin, 2014 (pp. 209–221). New York: Springer. doi: https://doi.org/10.1007/978-3-319-19977-1_15

Liang, L., Hayashi, K., & Yuan, K.-H. (2016). The goodness of sample loadings of principal component analysis in approximating to factor loadings with high dimensional data. In L. A. van der Ark, D. M. Bolt, W-C. Wang, J. A. Douglas, & M. Wiberg (Eds.), *Quantitative psychology research: The 80th annual meeting of the psychometric society*, Beijing, China, 2015 (pp. 199–211). New York: Springer. doi:https://doi.org/10.1007/978-3-319-38759-8_15

Pourahmadi, M. (2013). *High-dimensional covariance estimation.* New York: Wiley. doi: https://doi.org/10.1002/9781118573617

Schneeweiss, H. (1997). Factors and principal components in the near spherical case. *Multivariate Behavioral Research, 32*, 375–401. doi: https://doi.org/10.1207/s15327906mbr3204_4

Schneeweiss, H., & Mathes, H. (1995). Factor analysis and principal components. *Journal of Multivariate Analysis, 55*, 105–124. doi: https://doi.org/10.1006/jmva.1995.1069

Schott, J. R. (2005). *Matrix analysis for statistics* (2nd ed.). New York: Wiley.

ten Berge, J. M. F., Krijnen, W. P., Wansbeek, T., & Shapiro, A. (1999). Some new results on correlation-preserving factor scores prediction methods. *Linear Algebra and Its Applications, 289*, 311–318. doi: https://doi.org/10.1016/S0024-3795(97)10007-6

Thurstone, L. L. (1947). *Multiple factor analysis.* Chicago: University of Chicago Press. doi: https://doi.org/10.1086/396060

Tibshirani, R. (1996). Regression shrinkage and selection via the Lasso. *Journal of the Royal Statistical Society B, 58*, 267–288. doi: https://doi.org/10.1111/j.2517-6161.1996.tb02080.x

Tipping, M. E., & Bishop, C. M. (1999). Probabilistic principal component analysis. *Journal of the Royal Statistical Society B, 61*, 611–622. doi: https://doi.org/10.1111/1467-9868.00196

Ulyanov, V. V., Wakaki, H., & Fujikoshi, Y. (2006). Berry-Esseen bound for high dimensional asymptotic approximation of Wilks' Lambda distribution. *Statistics & Probability Letters, 76*, 1191–1200. doi: https://doi.org/10.1016/j.spl.2005.12.027

van Driel, O. P. (1978). On various causes of improper solutions in maximum likelihood factor analysis. *Psychometrika, 43*, 225–243. doi: https://doi.org/10.1007/BF02293865

Velicer, W. F., & Jackson, D. N. (1990). Component analysis versus common factor analysis: Some issues in selecting an appropriate procedure. *Multivariate Behavioral Research, 25*, 1–28. doi: https://doi.org/10.1207/s15327906mbr2501_1

Wakaki, H., Fujikoshi, Y., & Ulyanov, V. V. (2014).   Asymptotic expansions of the distributions of MANOVA test statistics when the dimension is large.   *Hiroshima Mathematical Journal, 44*, 247–259.   doi: https://doi.org/10.32917/hmj/1419619745

Yuan, K.-H., & Chan, W. (2008). Structural equation modeling with near singular covariance matrices. *Computational Statistics and Data Analysis, 52*, 4842–4828. doi: https://doi.org/10.1016/j.csda.2008.03.030

Yuan, K.-H., & Chan, W. (2016).   Structural equation modeling with unknown population distributions: Ridge generalized least squares.   *Structural Equation Modeling, 23*, 163–179.   doi: https://doi.org/10.1080/10705511.2015.1077335

Reliabilities with Ordered Response Categories Items

Seohyun Kim[1], Zhenqiu (Laura) Lu[2*], and Allan Cohen[2]

[1] The University of Virginia, USA
[2] The University of Georgia, USA
zlu@uga.edu

Abstract. This study proposed a structural equation modeling (SEM) approach to test reliabilities with items having ordered response categories. The number of categories in the test can be equal or unequal for all items. A simulation study was conducted to evaluate the performance of this proposed reliability, and compare it with the popular Cronbach's alpha. Simulation conditions include factor structure, the numbers of ordered categories, and the distribution of thresholds of underlying continuous scores. Results indicated that the proposed reliability coefficient performed better than other reliabilities in most conditions. Conclusions and discussion were also provided.

Keywords: Reliability · Structural equation modeling · Categorical data.

DOI: 10.35566/isdsa2019c10

1 Introduction

Reliabilities have been widely used in social and behavioral sciences (Bollen, 1989). Different estimation approaches have been proposed. Among them, Cronbach's coefficient alpha (Cronbach, 1951) is probably the most commonly used reliability. It takes a lower bound of the internal consistency of the test (Green, Lissitz, & Mulaik, 1977; Novick & Lewis, 1967; Sijtsma, 2009). The assumptions to derive alpha include errors in observed item scores are not correlated, and items in a test are essentially τ-equivalent (Novick & Lewis, 1967; Raykov, 1997; Zhang & Yuan, 2016).

Items with ordered response categories are also very common (Finney & DiStefano, 2006). For such tests, there are a number of studies that investigated the effects of the number of response categories on Cronbach's alpha (e.g., Bandalos & Enders, 1996; Komorita & Graham, 1965; Lissitz & Green, 1975; Lozano, García-Cueto, & Muñiz, 2008; Weng, 2004). These studies showed that coefficient alpha generally increases as the number of response categories increases. It is not clear, however, what happens when there are mixed numbers of ordered categories on the same test, e.g., two categories for some items, four or five categories for other items.

One approach to estimating reliability is the structural equation modeling (SEM) (Bentler, 2009; Bollen, 1989; Green & Yang, 2009; Miller, 1995; Raykov,

1997; Raykov & Shrout, 2002). The SEM approach introduces factorial structure to the test when estimating reliability, so it is especially useful when estimating reliability for tests having latent clusters (Cho & Kim, 2015; Green & Yang, 2015; Raykov & Shrout, 2002; Yang & Green, 2010), in which the tests are assumed to have general factors (e.g., reading ability, math ability, or personality). With a standard linear SEM approach, item scores may be presented by using a CFA model (e.g., Raykov & Shrout, 2002; Yang & Green, 2011), and reliability is calculated as the ratio of true score variance to observed score variance, in which the true score variance is estimated from the CFA model. This type of reliability is also referred to as coefficient omega (McDonald, 1985; McDonald, 1999). Green and Yang (2009) proposed a nonlinear SEM reliability when items in a test have equal number of ordinal categories. However, in reality, tests often consist of items with different numbers of categories. For example, the PARCC assessment for mathematics (2016) and the Smarter Balanced assessments (2016) include items have from two to seven points and two to four points, respectively. Likewise, a test such as the Early Development Instrument (Janus & Offord, 2007) for measuring children's school readiness also has items scored with four to eight points.

The present study was conducted to examine the effects on reliability when the items in a test have equal or unequal numbers of categories. In this study, we focus on total test scores, i.e., the sum scores of all items in a test. This approach is more general than Green and Yang (2009) because it considers both cases: items with the equal numbers of ordered categories and unequal numbers of ordered categories. We first provide a numerical formula to calculate the nonlinear SEM reliability, and then conduct a simulation study to evaluate the performance of this nonlinear SEM reliability by comparing the reliability coefficient to coefficient alpha and population reliability under different conditions. Based on a simulation study, conclusions and discussion are provided.

2   Reliability in Classical Test Theory

Suppose there are J items in a test. In classical test theory (CTT), an observed score X_j on item j ($j = 1, \ldots, J$) is composed of two components, a true score, T_j, and an error score, ϵ_j:

$$X_j = T_j + \epsilon_j. \tag{1}$$

Let X, T and ϵ be the sum of observed scores, of true scores, and of error scores, respectively, across J items. Then $X = \sum_{j=1}^{J} X_j$, $T = \sum_{j=1}^{J} T_j$, $\epsilon = \sum_{j=1}^{J} \epsilon_j$, and $X = T + \epsilon$. The reliability coefficient of a test is defined as

$$\rho = \frac{\sigma_T^2}{\sigma_X^2}, \tag{2}$$

where σ_T^2 is the variance of T, and σ_X^2 is the variance of X (Lord & Novick, 1968).

Coefficient alpha (Cronbach, 1951) is often used to estimate reliability

$$\alpha = \frac{J}{J-1}\left(\frac{\Sigma_{j \neq j'} Cov(X_j, X_{j'})}{\sigma_X^2}\right), \tag{3}$$

where $Cov(X_j, X_{j'})$ is the covariance between observed scores for items j and j'. When test items are essentially τ-equivalent, coefficient alpha is equal to ρ (Novick & Lewis, 1967).

3   SEM Approach to Reliability

In the framework of SEM, item scores can be represented as follows (Green & Yang, 2009; Raykov & Shrout, 2002):

$$X_j^* = \lambda_{1j}\eta_1 + \lambda_{2j}\eta_2 + \cdots + \lambda_{Mj}\eta_M + e_j, \tag{4}$$

where X_j^* is a continuous score for item j, M is the number of latent factors, η_m $(m = 1, \dots, M)$ are latent factors weighted by corresponding factor loadings λ_{mj}, and e_j is a measurement error term. On the right side of Equation (4), $\sum_{m=1}^{M} \lambda_{mj}\eta_m$ can be considered as a true score and e_j can be considered as an error score in CTT. When X_j^* $(j = 1, \dots, J)$ are continuous, parameters in Equation (4) are typically estimated by maximum likelihood estimation (MLE) with the assumption that observed variables are normally distributed (Bollen, 1989). Suppose X^* and T are the sum of observed scores and of true scores, respectively. Then $X^* = \sum_{j=1}^{J} X_j^*$ and $T = \sum_{j=1}^{J} \sum_{m=1}^{M} \lambda_{mj}\eta_m$. The linear SEM reliability ρ_{lin} is calculated as the ratio of true sum score variance to observed sum score variance as in Equation (2):

$$\rho_{lin} = \frac{\sigma_T^2}{\sigma_{X^*}^2} = \frac{Var(\sum_{j=1}^{J} \sum_{m=1}^{M} \lambda_{mj}\eta_m)}{Var(\sum_{j=1}^{J} X_j^*)}. \tag{5}$$

In this case, ρ_{lin} measures the proportion of observed sum score variance that is attributed to the latent factors, $\eta_1, \eta_2, \dots, \eta_M$ (Bollen, 1989). This linear SEM reliability, ρ_{lin}, is also equivalent to coefficient omega (McDonald, 1985; McDonald, 1999).

When the observed data are ordinal categorical, fitting linear SEM models such as Equation (4) using the MLE method is not desirable as categorical data violate the assumption of the MLE method, and the MLE provides inflated chi-square estimates and attenuated factor loadings (Bollen, 1989). To address this problem, we consider the observed categorical scores (X_j) as produced from underlying continuous variables (X_j^*), and that a linear SEM model holds for X_j^* as in Equation (4) (Bollen, 1989; Finney & DiStefano, 2006). We assume a nonlinear relationship between X_j and X_j^*:

$$X_j = \begin{cases} C_j - 1, \text{ if } X_j^* > v_{C_j-1} \\ \vdots \quad\quad\quad \vdots \\ 1, \quad \text{ if } v_1 < X_j^* \leq v_2 \\ 0, \quad \text{ if } \quad X_j^* \leq v_1 \end{cases}, \tag{6}$$

where C_j is the number of categories for X_j, and the ν_i ($i = 1, 2, \ldots, C_j - 1$) are the category thresholds. This allows for different numbers of ordered score categories for each item j. If X_j^* is less than ν_1, X_j is equal to 0, for $v_1 < X_j^* \leq v_2$, X_j is equal to 1, and if X_j^* is above ν_{C_j-1}, X_j is equal to $C_j - 1$.

In order to estimate the reliability for the nonlinear measurement model, we use the correlation between two parallel tests:

$$\rho_{X\tilde{X}} = \frac{Cov\left(X, \tilde{X}\right)}{\sqrt{Var\left(X\right) Var\left(\tilde{X}\right)}}, \tag{7}$$

where X and $\tilde{X}$ are observed sum scores across J items from two parallel tests, which indicates (a) the same latent factors affect the items from the two tests, and (b) the variance of errors and factor loadings are the same for corresponding items for both tests. For categorical items on parallel tests, the thresholds are the same for corresponding items (Bollen, 1989; Green & Yang, 2009). For example, for item j, the underlying continuous scores from two parallel tests using a confirmatory factor analysis model with M latent factors can be expressed as follows:

$$\begin{cases} X_j^* = \lambda_{1j}\eta_1 + \lambda_{2j}\eta_2 + \cdots + \lambda_{Mj}\eta_M + e_j \\ \tilde{X}_j^* = \lambda_{1j}\eta_1 + \lambda_{2j}\eta_2 + \cdots + \lambda_{Mj}\eta_M + \tilde{e}_j \end{cases}, \tag{8}$$

where λ_{mj} is a factor loading for factor η_m, e_j and $\tilde{e}_j$ are error terms for X_j^* and $\tilde{X}_j^*$, respectively.

The numerator of $\rho_{X\tilde{X}}$ in Equation (7) is a covariance between sum scores and can be presented as a function of covariances between two items from the parallel tests:

$$Cov\left(X, \tilde{X}\right) = Cov\left(\sum_{j=1}^{J} X_j, \sum_{j'=1}^{J} \tilde{X}_{j'}\right) = \sum_{j=1}^{J}\sum_{j'=1}^{J} Cov(X_j, \tilde{X}_{j'}). \tag{9}$$

Suppose item j has C_j categories, item j' has $C_{j'}$ categories, and the vectors of underlying continuous variables $\boldsymbol{X}^* = (X_1^*, \ldots, X_J^*)$ and $\widetilde{\boldsymbol{X}}^* = \left(\tilde{X}_1^*, \ldots, \tilde{X}_J^*\right)$ follow a multivariate normal distribution with variances of 1. Appendix A shows that $Cov\left(X_j, \tilde{X}_{j'}\right)$ can be represented as

$$Cov\left(X_j, \tilde{X}_{j'}\right) = \sum_{k=1}^{C_j-1}\sum_{l=1}^{C_{j'}-1} \Phi_2\left(\nu_{j_k}, h_{j'_l}; \rho_{X_j^*X_{j'}^*}\right) - \sum_{k=1}^{C_j-1} \Phi_1\left(\nu_{j_k}\right) \sum_{l=1}^{C_{j'}-1} \Phi_1\left(h_{j'_l}\right),$$

$$\tag{10}$$

where $\{v_{j_1}, \ldots, v_{j_{C_j-1}}\}$ and $\{h_{j'_1}, \ldots, h_{j'_{C_{j'}-1}}\}$ are thresholds for items j and j', respectively, $\Phi_1\left(\nu_{j_k}\right)$ is the cumulative univariate normal distribution function of threshold ν_{j_k}, $\Phi_2\left(\nu_{j_k}, h_{j'_l}; \rho_{X_j^*X_{j'}^*}\right)$ is the cumulative bivariate normal distribution function of thresholds ν_{j_k} and $h_{j'_l}$ with correlation $\rho_{X_j^*X_{j'}^*}$, where $\rho_{X_j^*X_{j'}^*}$

is the correlation between two underlying continuous variables. If X_j^* and $X_{j'}^*$ consist of M latent factors as in Equation (8), then $\rho_{X_j^* X_{j'}^*}$ can be represented as

$$\rho_{X_j^* X_{j'}^*} = \sum_{m=1}^{M} \sum_{m'=1}^{M} \lambda_{mj} \lambda_{m'j'} \rho_{\eta_m \eta_{m'}}, \tag{11}$$

where $\rho_{\eta_m \eta_{m'}}$ is the correlation between η_m and $\eta_{m'}$. We specifically indicate this correlation derived from a model as $\rho_{M_{jj'}}$ to avoid confusion.

Because of parallel tests X_j and $\tilde{X}_j$, the denominator of $\rho_{X\tilde{X}}$ in Equation (7) becomes

$$\begin{aligned}
Var(X) = Var\left(\sum_{j=1}^{J} X_j\right) &= \sum_{j=1}^{J} \sum_{j'=1}^{J} Cov(X_j, X_{j'}) \\
&= \sum_{j=1}^{J} \sum_{j'=1}^{J} \left(\sum_{k=1}^{C_j-1} \sum_{l=1}^{C_{j'}-1} \Phi_2\left(\nu_{j_k}, h_{j'_l}; \rho_{X_j^* X_{j'}^*}\right)\right. \\
&\quad \left. - \sum_{k=1}^{C_j-1} \Phi_1(\nu_{j_k}) \sum_{l=1}^{C_{j'}-1} \Phi_1\left(h_{j'_l}\right)\right).
\end{aligned} \tag{12}$$

Therefore, the nonlinear reliability coefficient using the SEM framework is expressed as

$$\rho_{non} = \frac{A}{B}. \tag{13}$$

where

$$\begin{aligned}
A = \sum_{j=1}^{J} \sum_{j'=1}^{J} &\left[\sum_{k=1}^{C_j-1} \sum_{l=1}^{C_{j'}-1} \Phi_2\left(\nu_{j_k}, h_{j'_l}; \rho_{M_{jj'}}\right)\right. \\
&\left. - \sum_{k=1}^{C_j-1} \Phi_1(\nu_{j_k}) \sum_{l=1}^{C_{j'}-1} \Phi_1\left(h_{j'_l}\right)\right],
\end{aligned}$$

and

$$\begin{aligned}
B = \sum_{j=1}^{J} \sum_{j'=1}^{J} &\left[\sum_{k=1}^{C_j-1} \sum_{l=1}^{C_{j'}-1} \Phi_2\left(\nu_{j_k}, h_{j'_l}; \rho_{X_j^* X_{j'}^*}\right)\right. \\
&\left. - \sum_{k=1}^{C_j-1} \Phi_1(\nu_{j_k}) \sum_{l=1}^{C_{j'}-1} \Phi_1\left(h_{j'_l}\right)\right].
\end{aligned}$$

This can be seen as an extension of Equation 21 in Green and Yang (2009). From equation (13), we can estimate the internal consistency reliability of a test consisting of items with different numbers of ordered response categories by fitting the data using a nonlinear SEM model and replacing parameters in Equation (13) with corresponding estimates.

4   Simulation Study

In this section, a simulation study is presented to investigate the performance of the proposed nonlinear SEM reliability ρ_{non}. We assumed the data are from a one-factor model and the numbers of item response categories in the test can be different or the same. For comparison purposes, coefficient alpha (α) and population reliabilities for observed sum scores ($\rho_{X\bar{X}}$) were also presented.

4.1   Simulation Study Design

Factor structure. Models 1 and 2 are set up based on a one-factor model. Model 1 has 9 items and Model 2 has 18 items. Factor loadings (λ) were all set to 0.7. Errors were assumed independent.

Numbers of ordered categories. The number of item response categories in a test was set either the same (either all two categories or all five categories) or different (a combination of two categories and five categories). Tests with either all two or all five ordered categories were labeled as conditions C2 and C5 in the simulation, respectively. Tests with a combination of two- and five-category items were labeled as condition C25, in which every third item has five ordered categories and the rest items have two ordered categories.

Distribution of underlying continuous variables. The underlying continuous variable $\boldsymbol{X}^* = (X_1^*, \ldots, X_J^*)$ was assumed to follow a multivariate normal distribution: $\boldsymbol{X}^* \sim N(\boldsymbol{0}, \boldsymbol{\Sigma})$, where J is the number of items, $\boldsymbol{0}$ is a $J \times 1$ vector of 0, and $\boldsymbol{\Sigma}$ is the $J \times J$ covariance matrix of $\boldsymbol{X}^*$ with diagonal elements of 1. The underlying continuous score on item j for examinee i, x_{ij}^*, was generated to have a variance of one (Bandalos & Enders, 1996; Bernstein & Teng, 1989; Flora & Curran, 2004; Green & Yang, 2009; Yang & Green, 2015; Yuan & Lu, 2008)

$$x_{ij}^* = \boldsymbol{\lambda}_j^T \boldsymbol{\eta}_i + \sqrt{\left(1 - \boldsymbol{\lambda}_j^T Var\left(\boldsymbol{\eta}_i\right) \boldsymbol{\lambda}_j\right)}\epsilon_{ij}, \tag{14}$$

where $\boldsymbol{\lambda}_j$ is a vector of factor loadings for item j, $\boldsymbol{\eta}_i$ is a vector of latent factors, ϵ_{ij} is an error score with a mean of 0 and a variance of 1, and ϵ_{ij} is uncorrelated with $\boldsymbol{\eta}_i$. Off diagonal elements in the covariance matrix $\boldsymbol{\Sigma}$ were determined by Equation (14).

Thresholds. To transform the underlying continuous variables to categorical data, thresholds are set for different conditions: symmetric, moderate skewed, and mixed skewed. (a) In symmetric cases, threshold for two response categories was set as {0}, and the thresholds for five-categories were {-1.645, -0.643, 0.643, 1.645} (Muthén & Kaplan, 1985). (b) In moderate skewed cases, {0.7} and {-0.050, 0.772, 1.341, 1.881} were used to generate two- and five-categories (Muthén & Kaplan, 1985), respectively. These two conditions have the same skewness 1.2 (Doane & Seward, 2011). (c) In the mixed skewed case, every third item had negative moderate skewed thresholds and the rest items have positive moderate skewed thresholds.

4.2 Data Generation

In total there were 18 conditions: $18 = 2$ factor structures $\times$ 3 sets of item response categories $\times$ 3 distributions of thresholds. Under each condition, we first generated population data consisting of 100,000 observations. And then, we randomly sampled 500 observations without replacement. Replication number is 100.

In order to calculate the population reliability, $\rho_{X\tilde{X}}$, we generated another set of parallel population data. The correlation of the two observed sum scores from these two parallel tests was treated as the population reliability.

4.3 Data Analysis

A one-factor model was fit to the corresponding generated data. Under each condition, ρ_{non} and α were obtained and compared. The proposed ρ_{non} was calculated as in Equation (13) by using R (R Core Team, 2017). Model parameters were estimated by weighted least squares with mean and variance adjustment (WLSMV; Muthén, du Toit, & Spisic, 1997; Muthén & Muthén, 1998-2015) in Mplus 7.4 (Muthén & Muthén, 1998-2015). Cronbach's α was obtained by using the R package psych (Revelle, 2017). Polychoric correlations were also estimated using Mplus 7.4.

5 Simulation Study Results

We first examined whether models converged. Converged cases were used to calculate ρ_{non} and α. Table 1 lists the means of reliability coefficients.

Table 1 shows the C2 condition had the lowest $\rho_{X\tilde{X}}$ and ρ_{non}, ranging from 0.77 to 0.90. The C5 condition had the highest $\rho_{X\tilde{X}}$ and ρ_{non}, ranging from 0.86 to 0.94. The two reliability coefficients for the mixed number of categories conditions (i.e., condition C25) ranged from 0.80 to 0.92, and were between those for C2 and C5. The values of ρ_{non} were all close to the values of $\rho_{X\tilde{X}}$, and the standard deviations of ρ_{non} were all less than or equal to 0.02. For some conditions, ρ_{non} was .01 higher than $\rho_{X\tilde{X}}$. These results are consistent with the results in Yang and Green (2015) with items having the same number of ordered response categories.

Cronbach's α in Table 1 had the lowest values, when the category condition was either C2 or C25, and had the highest values for the C5 condition. Coefficient α values were nearly identical with the corresponding $\rho_{X\tilde{X}}$ when items had the same number of response categories (i.e., conditions C2 and C5), but it was slightly lower than $\rho_{X\tilde{X}}$ when the test included items with uneven numbers of response categories (i.e., condition C25) and the responses had skewed distributions. When the observed scores were generated from the model with nine items with an uneven number of categories (i.e., Model 1 with C25) and the mixed skew condition, the difference between α and the $\rho_{X\tilde{X}}$ was 0.05. The standard deviations of α were less than or equal to 0.02.

Factor models fit the corresponding data well under each condition: across all replications the CFI were all higher than .98, and the RMSEA were all smaller than .06 with mean values for each condition ranging from .01 to .02.

6   Conclusions and Discussion

A generalized nonlinear SEM approach was proposed in this study to estimate internal consistency reliability when test items have ordered categories. The numbers of categories of items in the test can be equal or unequal. A simulation study was conducted to evaluate the performance of this coefficient by comparing it with Cronbach's alpha and population reliability of observed sum scores.

Results showed that the proposed nonlinear SEM reliability estimate, ρ_{non}, was very close to the population reliability $\rho_{X\tilde{X}}$ values across all the conditions. And the more response categories, the more information, the better performance, even in the mixed-category cases. The differences between these two coefficients were mostly less than 0.01. ρ_{non} and $\rho_{X\tilde{X}}$ both had the highest values under the C5 condition, and had lowest values under the C2 condition. Both reliability coefficients for the C25 condition were placed between the reliability coefficients for C2 and C5. For some conditions, the values of ρ_{non} were slightly higher than the values of $\rho_{X\tilde{X}}$. Results of this study also suggest that this general approach is flexible because it can be used for the test with items having equal number of ordered categories or unequal number of ordered categories.

Regarding Cronbach's alpha, it performed well only when the numbers of categories are equal (i.e., conditions C2 and C5) in the simulation and when the observed categorical scores were essentially τ-equivalent. The Cronbach's α was lower than the population reliability $\rho_{X\tilde{X}}$ when items had mixed numbers of response categories or when the scores were skewed. In reality, tests often have items with different numbers of categories, and different thresholds may result in non-essentially τ-equivalent tests. In these cases, Cronbach's alpha is not recommended.

The simulation in this study was conducted under the conditions of correctly specified factor models. When the factor model is mis-specified, the nonlinear SEM reliability estimate may be biased. The mis-specification can lead to incorrect parameter estimation, resulting in incorrect nonlinear reliability estimates. Also, the simulation study assumed the distribution of the latent variables was multivariate normality. So this study can be further extended to check the impact of model misspecification and the robustness when these assumptions are violated.

Table 1. The Means of Reliability Coefficients and their Standard Deviations

Model	Reliability	Normal			Moderate skewness			Mixed skewness		
		C2	C5	C25	C2	C5	C25	C2	C5	C25
1	$\rho_{X\bar{X}}$	0.81	0.87	0.83	0.79	0.87	0.81	0.77	0.86	0.80
	ρ_{non}	0.82	0.88	0.84	0.80	0.87	0.82	0.79	0.86	0.80
		(0.01)	(0.02)	(0.01)	(0.01)	(0.01)	(0.01)	(0.01)	(0.01)	(0.01)
	α	0.81	0.87	0.82	0.79	0.87	0.78	0.76	0.85	0.75
		(0.01)	(0.01)	(0.01)	(0.02)	(0.01)	(0.01)	(0.01)	(0.01)	(0.01)
2	$\rho_{X\bar{X}}$	0.90	0.93	0.91	0.88	0.93	0.90	0.87	0.92	0.89
	ρ_{non}	0.90	0.94	0.92	0.89	0.93	0.90	0.88	0.93	0.89
		(0.01)	(0.00)	(0.01)	(0.01)	(0.01)	(0.01)	(0.01)	(0.00)	(0.01)
	α	0.90	0.93	0.90	0.88	0.93	0.88	0.87	0.92	0.87
		(0.01)	(0.00)	(0.01)	(0.01)	(0.01)	(0.01)	(0.01)	(0.00)	(0.01)

Note:
- *Reliability $\rho_{X\bar{X}}$: population reliability for observed sum scores*
- *Reliability ρ_{non} : the proposed nonlinear SEM reliability*
- *Reliability α : Cronbach's alpha*
- *C: condition for the number of response categories, C2: all items have 2 response categories,*
- *C5: all items have 5 response categories, C25: items have a combination of 2- and 5- response categories.*

References

Bandalos, D. L., & Enders, C. K. (1996). The effects of nonnormality and number of response categories on reliability. *Applied Measurement in Education*, 9(2), 151-160.

Bentler, P. M. (2009). Alpha, dimension-free, and model-based internal consistency reliability. *Psychometrika*, 74(1), 137-143.

Bernstein, I. H., & Teng, G. (1989). Factoring items and factoring scales are different: Spurious evidence for multidimensionality due to item categorization. *Psychological Bulletin*, 105(3), 467-477.

Bollen, K. A. (1989) *Structural equations with latent variables*. New York, NY: Wiley.

Cho, E., & Kim, S. (2015). Cronbach's coefficient alpha: Well known but poorly understood. *Organizational Research Methods*, 18(2), 207-230.

Cronbach, L. J. (1951). Coefficient alpha and the internal structure of tests. *Psychometrika*, 16(3), 297-334.

Cronbach, L. J., & Shavelson, R. J. (2004). My current thoughts on coefficient alpha and successor procedures. *Educational and psychological measurement*, 64(3), 391-418.

Doane, D. P., & Seward, L. E. (2011). Measuring skewness: a forgotten statistic? *Journal of Statistics Education*, 19(2), 1-18.

Finney, S. J., & DiStefano, C. (2006). Non-normal and categorical data in structural equation modeling. In G. R. Hancock & R. O. Mueller (Eds.). *Structural equation modeling: A second course* (pp. 269-314). Greenwich: Information Age.

Flora, D. B., & Curran, P. J. (2004). An empirical evaluation of alternative methods of estimation for confirmatory factor analysis with ordinal data. *Psychological methods*, 9(4), 466-491.

Green, S. B., Lissitz, R. W., & Mulaik, S. A. (1977). Limitations of coefficient alpha as an index of test unidimensionality. *Educational and Psychological Measurement*, 37(4), 827-838.

Green, S. B., & Yang, Y. (2009). Reliability of summed item scores using structural equation modeling: An alternative to coefficient alpha. *Psychometrika*, 74(1), 155-167.

Green, S. B., & Yang, Y. (2015). Evaluation of dimensionality in the assessment of internal consistency reliability: Coefficient alpha and omega coefficients. *Educational Measurement: Issues and Practice*, 34(4), 14-20.

Janus, M., & Offord, D. R. (2007). Development and psychometric properties of the early development instrument (EDI): A measure of children's school readiness. *Canadian Journal of Behavioural Science*, 39, 1-22.

Komorita, S. S., & Graham, W. K. (1965). Number of scale points and the reliability of scales. *Educational and Psychological Measurement*, 25(4), 987-995.

Lissitz, R. W., & Green, S. B. (1975). Effect of the number of scale points on reliability: A Monte Carlo approach. *Journal of Applied Psychology*, 60(1), 10-13.

Lord, F. M., & Novick, M. R. (1968). *Statistical theories of mental test scores.* Reading, MA: Addison-Wesley.

Lozano, L. M., García-Cueto, E., & Muñiz, J. (2008). Effect of the number of response categories on the reliability and validity of rating scales. *Methodology*, 4(2), 73-79.

McDonald, R. P. (1985). *Factor analysis and related methods.* Hillsdale, NJ: Erlbaum.

McDonald, R. P. (1999) *Test theory: A unified treatment.* Mahwah, NJ: Erlbaum.

Miller, M. B. (1995). Coefficient alpha: A basic introduction from the perspectives of classical test theory and structural equation modeling. *Structural Equation Modeling*, 2(3), 255-273.

Muthén, B., du Toit, S. H. C., & Spisic, D. (1997). Robust inference using weighted least squares and quadratic estimating equations in latent variable modeling with categorical and continuous outcomes. *Unpublished manuscript*

Muthén, B., & Kaplan, D. (1985). A comparison of some methodologies for the factor analysis of non-normal Likert variables. *British Journal of Mathematical and Statistical Psychology*, 38(2), 171-189.

Muthén, L. K., & Muthén, B. O. (1998-2015). *Mplus user's guide* (7th ed.). Los Angeles, CA: Muthén & Muthén.

Novick, M., & Lewis, C. (1967). Coefficient alpha and the reliability of composite measurements. *Psychometrika*, 32(1), 1-13.

Partnership for Assessment of Readiness for College and Careers (PARCC) (2016). *Final Technical Report for 2015 Administration*

R Core Team (2017). R: A language and environment for statistical computing. R Foundation for Statistical Computing, Vienna, Austria. URL= http://www.R-project.org/.

Raykov, T. (1997). Estimation of composite reliability for congeneric measures. *Applied Psychological Measurement*, 21(2), 173-184.

Raykov, T., & Shrout, P. E. (2002). Reliability of scales with general structure: Point and interval estimation using a structural equation modeling approach. *Structural equation modeling*, 9(2), 195-212.

Revelle, W. R. (2017). psych: Procedures for personality and psychological research. *R package version 1.8.4*

Sijtsma, K. (2009). On the use, the misuse, and the very limited usefulness of Cronbach's alpha. *Psychometrika*, 74(1), 107-120.

Smarter Balanced Assessment Consortium. (2016) 2013-14. *Technical Report.*

Weng, L. J. (2004). Impact of the number of response categories and anchor labels on coefficient alpha and test-retest reliability. *Educational and Psychological Measurement*, 64(6), 956-972.

Yang, Y., & Green, S. B. (2010). A note on structural equation modeling estimates of reliability. *Structural Equation Modeling*, 17(1), 66-81.

Yang, Y., & Green, S. B. (2011). Coefficient alpha: A reliability coefficient for the 21st century? *Journal of Psychoeducational Assessment*, 29(4), 377-392.

Yang, Y., & Green, S. B. (2015). Evaluation of structural equation modeling estimates of reliability for scales with ordered categorical items. *Methodology*, 11(1), 23-34.

Yuan, K.-H., & Lu, L., (2008). SEM with missing data and unknown population distributions using two-stage ML: Theory and its application. *Multivariate Behavioral Research*, 43(4), 621-652.

Zhang, Z. & Yuan, K.-H. (2016). Robust coefficients alpha and omega and confidence intervals with outlying observations and missing data: Methods and software. *Educational and Psychological Measurement*, 76(3), 387–411.

Appendix

Appendix A describes the computation of covariance between observed scores from two parallel tests. This covariance can be expressed using the expectations of X_j and $\tilde{X}_{j'}$ as follows:

$$Cov\left(X_j, \ \tilde{X}_{j'}\right) = E\left(X_j\tilde{X}_{j'}\right) - E\left(X_j\right)E\left(\tilde{X}_{j'}\right). \tag{15}$$

Suppose item j has C_j categories, and item j' has $C_{j'}$ categories. Note that the vectors of underlying continuous variables $\boldsymbol{X}^* = (X_1^*, \ldots, X_J^*)$ and $\widetilde{\boldsymbol{X}}^* = \left(\tilde{X}_1^*, \ldots, \tilde{X}_J^*\right)$ are assumed to follow a multivariate normal distribution with covariance matrix $\boldsymbol{\Sigma}$ that has 1 as the diagonal elements. The first term on the right-hand side of Equation (15) can be expressed as

$$E\left(X_j\tilde{X}_{j'}\right) = \sum_{k=0}^{C_j-1}\sum_{l=0}^{C_{j'}-1} kl P\left(v_{j_k} < X_j^* \le v_{j_{k+1}}, \quad h_{j'_l} < \tilde{X}_{j'}^* \le h_{j'_{l+1}}\right), \tag{16}$$

where $\{v_{j_0}, \ v_{j_1}, \ \ldots, v_{j_{C_j}}\}$ and $\{h_{j'_0}, \ h_{j'_1}, \ \ldots, h_{j'_{C_{j'}}}\}$ are thresholds for items j and j', respectively, and $v_{j_0}, h_{j'_0}, v_{j_{C_j}},$ and $h_{j'_{C_{j'}}}$ are $-\infty, -\infty, \infty,$ and $\infty,$ respectively. Equation (16) can be rewritten as

$$E\left(X_j \tilde{X}_{j'}\right) = \sum_{k=0}^{C_j-1} \sum_{l=0}^{C_{j'}-1} kl P\left(\nu_{j_k} < X_j^* \le \nu_{j_{k+1}}, \quad h_{j_l'} < \tilde{X}_{j'}^* \le h_{j_{l+1}'}\right)$$

$$= \sum_{k=0}^{C_j-1} \sum_{l=0}^{C_{j'}-1} kl \left[\Phi_2\left(\nu_{j_{k+1}}, h_{j_{l+1}'}; \rho_{X_j^* \tilde{X}_{j'}^*}\right) - \Phi_2\left(\nu_{j_{k+1}}, h_{j_l'}; \rho_{X_j^* \tilde{X}_{j'}^*}\right)\right.$$

$$\left. - \Phi_2\left(\nu_{j_k}, h_{j_{l+1}'}; \rho_{X_j^* \tilde{X}_{j'}^*}\right) + \Phi_2\left(\nu_{j_k}, h_{j_l'}; \rho_{X_j^* \tilde{X}_{j'}^*}\right)\right]$$

$$= \sum_{k=1}^{C_j-1} \sum_{l=1}^{C_{j'}-1} (kl - k - l + 1)\Phi_2\left(\nu_{j_k}, h_{j_l'}; \rho_{X_j^* \tilde{X}_{j'}^*}\right)$$

$$+ \sum_{k=1}^{C_j-1} (kC_{j'} - C_{j'} - k + 1)\Phi_2\left(\nu_{j_k}, h_{j_{C_{j'}}'}; \rho_{X_j^* \tilde{X}_{j'}^*}\right)$$

$$+ \sum_{l=1}^{C_{j'}-1} (C_j l - l - C_j + 1)\Phi_2\left(\nu_{j_{C_j}}, h_{j_l'}; \rho_{X_j^* \tilde{X}_{j'}^*}\right)$$

$$+ (C_j C_{j'} - C_j - C_{j'} + 1)\Phi_2\left(\nu_{j_{C_j}}, h_{j_{C_{j'}}'}; \rho_{X_j^* \tilde{X}_{j'}^*}\right)$$

$$- \left\{\sum_{k=1}^{C_j-1} \sum_{l=1}^{C_{j'}-1} (kl - l)\Phi_2\left(\nu_{j_k}, h_{j_l'}; \rho_{X_j^* \tilde{X}_{j'}^*}\right)\right.$$

$$\left. + \sum_{l=1}^{C_{j'}-1} (C_j l - l)\Phi_2\left(\nu_{j_{C_j}}, h_{j_l'}; \rho_{X_j^* \tilde{X}_{j'}^*}\right)\right\}$$

$$- \left\{\sum_{k=1}^{C_j-1} \sum_{l=1}^{C_{j'}-1} (kl - k)\Phi_2\left(\nu_{j_k}, h_{j_l'}; \rho_{X_j^* \tilde{X}_{j'}^*}\right)\right.$$

$$\left. + \sum_{k=1}^{C_j-1} (kC_{j'} - k)\Phi_2\left(\nu_{j_k}, h_{j_{C_{j'}}'}; \rho_{X_j^* \tilde{X}_{j'}^*}\right)\right\}$$

$$+ \left\{\sum_{k=1}^{C_j-1} \sum_{l=1}^{C_{j'}-1} kl\Phi_2\left(\nu_{j_k}, h_{j_l'}; \rho_{X_j^* \tilde{X}_{j'}^*}\right)\right\}$$

$$= \sum_{k=1}^{C_j-1} \sum_{l=1}^{C_{j'}-1} \Phi_2\left(\nu_{j_k}, h_{j_l'}; \rho_{X_j^* \tilde{X}_{j'}^*}\right) - (C_{j'} - 1)\sum_{k=1}^{C_j-1} \Phi_1\left(\nu_{j_k}\right)$$

$$- (C_j - 1)\sum_{l=1}^{C_{j'}-1} \Phi_1\left(h_{j_l'}\right) + (C_j - 1)(C_{j'} - 1),$$

where $\Phi_2\left(\nu_{j_k}, h_{j'_l}; \rho_{X^*_j \tilde{X}^*_{j'}}\right)$ is the cumulative bivariate normal distribution function of ν_{j_k} and $h_{j'_l}$ with correlation $\rho_{X^*_j \tilde{X}^*_{j'}}$, and $\Phi_1(\nu_{j_k})$ is the cumulative univariate normal distribution function of ν_{j_k}. Since X^*_j and $\tilde{X}^*_{j'}$ are the underlying continuous variables from the parallel items, $\rho_{X^*_j \tilde{X}^*_{j'}} = \rho_{X^*_j X^*_{j'}}$. Thus,

$$E\left(X_j \tilde{X}_{j'}\right) = \sum_{k=1}^{C_j-1} \sum_{l=1}^{C_{j'}-1} \Phi_2\left(\nu_{j_k}, h_{j'_l}; \rho_{X^*_j X^*_{j'}}\right) - (C_{j'} - 1) \sum_{k=1}^{C_j-1} \Phi_1\left(\nu_{j_k}\right).$$
$$- (C_j - 1) \sum_{l=1}^{C_{j'}-1} \Phi_1\left(h_{j'_l}\right) + (C_j - 1)(C_{j'} - 1)$$

Similarly, the expectation of X_j can be represented as

$$E(X_j) = \sum_{k=0}^{C_j-1} kP(X_j = k) = \sum_{k=0}^{C_j-1} kP\left(\nu_{j_k} < X^*_j \leq \nu_{j_{k+1}}\right)$$
$$= \sum_{k=0}^{C_j-1} k\left[\Phi_1\left(\nu_{j_{k+1}}\right) - \Phi_1\left(\nu_{j_k}\right)\right]$$
$$= \sum_{k=1}^{C_j} (k-1)\Phi_1\left(\nu_{j_k}\right) - \sum_{k=0}^{C_j-1} k\Phi_1\left(\nu_{j_k}\right)$$
$$= \sum_{k=1}^{C_j-1} (k-1)\Phi_1\left(\nu_{j_k}\right) - \sum_{k=1}^{C_j-1} k\Phi_1\left(\nu_{j_k}\right) + (C_j - 1)\Phi_1\left(\nu_{j_{C_j}}\right)$$
$$= -\sum_{k=1}^{C_j-1} \Phi_1\left(\nu_{j_k}\right) + (C_j - 1).$$

Thus,

$$E(X_j) E\left(\tilde{X}_{j'}\right) = \sum_{k=1}^{C_j-1} \Phi_1\left(\nu_{j_k}\right) \sum_{l=1}^{C_{j'}-1} \Phi_1\left(h_{j'_l}\right) - (C_j - 1) \sum_{l=1}^{C_{j'}-1} \Phi_1\left(h_{j'_l}\right).$$
$$- (C_{j'} - 1) \sum_{k=1}^{C_j-1} \Phi_1\left(\nu_{j_k}\right) + (C_j - 1)(C_{j'} - 1)$$

Therefore,

$$Cov\left(X_j, \tilde{X}_{j'}\right) = E\left(X_j \tilde{X}_{j'}\right) - E(X_j) E\left(\tilde{X}_{j'}\right).$$
$$= \sum_{k=1}^{C_j-1} \sum_{l=1}^{C_{j'}-1} \Phi_2\left(\nu_{j_k}, h_{j'_l}; \rho_{X^*_j X^*_{j'}}\right) - \sum_{k=1}^{C_j-1} \Phi_1\left(\nu_{j_k}\right) \sum_{l=1}^{C_{j'}-1} \Phi_1\left(h_{j'_l}\right)$$

More Accurate Estimators of Multiple Correlation Coefficient[*]

Bingjiang Li[1], Lu Peng[1], Kentaro Hayashi[2], and Ke-Hai Yuan[1,3]

[1] Nanjing University of Posts and Telecommunications, China
1218084111@njupt.edu.cn
[2] University of Hawaii at Manoa, USA
[3] University of Notre Dame, USA

Abstract. The squared multiple correlation (R^2) is commonly used to measure how well the outcome variable is linearly related to a set of predictors. Unfortunately, R^2 is biased for its population counterpart (ρ^2), and the bias increases as the number of variables (p) increases. Efforts have been made to modify R^2. The most notable result is the adjusted R^2 (R^2_{adj}), which incorporates the influence of the sample size (N) and p. However, R^2_{adj} is still biased, and an unbiased estimator of ρ^2 does not exist. Using empirical modeling and statistical learning, this article develops new formulas for estimating the population ρ. The development involves obtaining formulas for the empirical bias of R via Monte Carlo simulation across many conditions. Values of the empirical bias are then predicted by functions of N, p and the observed values of the R. Best-subset regression are used to identify the best predictors for the empirical bias. Improved formulas for estimating ρ are obtained via a bias correction to R. Results of cross validation show that empirically corrected estimators contain little bias and perform better than both R and R_{adj} in mean squared error and variance.

Keywords: Empirical modeling · Monte Carlo simulation · Bias correction · Best-subset regression.

DOI: 10.35566/isdsa2019c11

1 Introduction

Data in behavioral and social sciences often contain a large number of variables. Linear regression models are commonly used to analyze such data to investigate the linear relationship between an outcome variable and a set of predictors. The coefficient of determination, R^2, has been used as a measure for the strength of the linear relationship and its value is routinely reported in publication and presentation. For a regression model with an intercept, R^2 can be defined as the squared Pearson correlation coefficient between the predicted and the observed

[*] Supported by the National Natural Science Foundation of China (Grant No. 31971029).

values of the outcome variable. However, in estimating the regression coefficients for a given data set by e.g., the least squares (LS) method, sampling variability affects both parameter estimates and the value of R^2 (Cohen & Cohen, 1983; Brooks & Stevens, 1994; Wherry, 1931). More concretely, sampling error causes R^2 to be systematically higher than the corresponding population counterpart ρ^2 and, thus, R^2 is a positively biased estimator of ρ^2 (Carter, 1979; Larson, 1931; Wherry, 1931; Yin & Fan, 2001). In fact, because the value of R^2 increases even when predictors that have no relationship with the outcome variable are added to the model, it is sometimes misleading to make a conclusion only based on R^2. Cohen and Cohen (1983, p. 105) stated that "Although we may determine from a sample R^2 that the population R^2 is not likely to be zero, it is nevertheless not true that the sample R^2 is a good estimate of the population R^2."

Efforts have been made to modify R^2 to be closer to ρ^2. There are two main approaches: analytical and empirical methods. Analytical methods adjust the statistical bias to yield an corrected sample R^2. Over the decades, a variety of corrected formulas have been developed to shrink R^2 to obtain a less biased estimator (e.g., Browne, 1975; Darlington, 1968; Ezekiel, 1929; Lord, 1950; Nicholson, 1948; Stein, 1960; Wherry, 1931). Empirical methods estimate the average predictive power of a sample regression equation using independent samples. Typical empirical methods for this purpose are data splitting, cross-validation, multi-cross-validation, jackknife, and bootstrap (Ayabe, 1985; Cummings, 1982; Kromrey & Hines, 1995; Krus & Fuller, 1982). However, in providing recommendation for shrinkage estimates of ρ^2, some authors often do not distinguish the difference between the squared population correlation ρ^2 and the squared population cross-validation coefficient ρ_c^2. It is important to distinguish these two parameters because a reasonable shrinkage to one estimator does not necessarily mean the same shrinkage is suitable for other estimators. Kromrey and Hines (1995) demonstrate the inaccuracy of using empirical estimates of shrinkage (e.g., cross-validation, multi-cross-validation, jackknife, and bootstrap) to estimate the population squared multiple correlation ρ^2.

Though many corrections to R^2 have been proposed using analytic methods, none of them is unbiased for ρ^2. In this article, we will study the roles of N, p and the value of R in predicting the bias in R^2. For such a purpose, we will use the methods of empirical modeling and statistical learning as introduced in Yuan, Tian, and Yanagihara (2015). In particular, we aim to (a) develop a formula of correcting R^2; (b) identify the most reliable predictors of the causes of the bias in R^2 with normally distributed data; and (c) evaluate the performance of our corrected R^2 by comparing it with the most widely used adjusted R^2. Results show that our corrected R^2 is more accurate than the adjusted R^2. The idea of using statistical learning and empirical modeling can also be used to obtain a better estimator of ρ^2 under more complicated conditions.

2   Methods

The linear regression model is among the most commonly used statistical models across disciplines. With p predictors, the model can be written as

$$y = \beta_0 + \beta_1 x_1 + \beta_2 x_2 + \ldots + \beta_p x_p + \varepsilon, \tag{1}$$

where y is the outcome variable, x_j ($j = 1, 2, \ldots, p$) are predictors, ε is normally distributed with mean 0 and variance σ^2, and the β_js are regression coefficients. The population coefficient of determination for the model in equation (1) is

$$\rho^2 = 1 - \sigma^2 / \text{Var}(y). \tag{2}$$

To estimate the population coefficient of determination, researchers have developed a variety of forms for different types of models. Let $(x_{i1}, x_{i2}, \ldots, x_{ip}, y_i)$, $i = 1, 2, \ldots, N$ be the observed sample. For a linear regression model with an intercept term and parameters estimated by the LS method, Tarald (1985; see also Kvålseth, 1985) recommended

$$R_1^2 = 1 - \sum_{i=1}^{N}(y_i - \hat{y}_i)^2 / \sum_{i=1}^{N}(y_i - \bar{y})^2 \tag{3}$$

for estimating ρ^2, where $\hat{y}_i$ is the model implied value of y for the ith case. R_1^2 can be interpreted as the proportion of the total variance of y that is explained by the fitted model. For the linear model in (1), the R_1^2 is numerically equal to the squared sample Pearson correlation coefficient between y and the predicted value $\hat{y}$, which can also be calculated by

$$R^2 = \frac{\mathbf{s}_{yx}\mathbf{S}_{xx}^{-1}\mathbf{s}_{xy}}{s_{yy}}, \tag{4}$$

where s_{yy} is the sample variance of y, $\mathbf{S}_{xx}$ is the sample covariance matrix of $\mathbf{x} = (x_1, \ldots, x_p)'$ and $\mathbf{s}_{xy}$ is a vector of the sample covariances between $\mathbf{x}$ and y. Consequently, R_1^2 is commonly denoted as R^2. However, it is known that R^2 is positively biased for ρ^2. Because of the bias, researchers often prefer adjusting R^2 for the appropriate degrees of freedom. The most common adjustment for R^2 is to divide its numerator and denominator on the right side of equation (3) by their respective degrees of freedom so that the adjusted R^2 is defined as

$$R_{adj}^2 = 1 - a \sum_{i=1}^{N}(y_i - \hat{y}_i)^2 / \sum_{i=1}^{N}(y_i - \bar{y})^2, \tag{5}$$

where $a = (N - 1)/(N - p - 1)$ with N being the sample size, and p being the number of predictors. Unfortunately, although R^2 and R_{adj}^2 are widely used in measuring the accounted variance in multiple regression, neither of them is unbiased for ρ^2. The bias in R^2 is a function of both N and p in the regression equation (Huberty & Mourad, 1980). As we shall see, the size of the bias is also related to the value of R^2. In the development, we will consider many different functions and transformations of N, p and the value of R to predict the bias in R^2 using statistical learning. Because R^2 and R are a one-to-one transformation, we will work with correcting the bias in R. Then we can use Fisher z transformation to stabilize the variance of R so that we are able to further model and predict the bias in R using the LS method.

Let $(x_{i1}, x_{i2}, \ldots, x_{ip}, y_i)$, $i = 1, 2, \ldots, N$ be a random sample from a multi-variate normal distribution $N(\boldsymbol{\mu}, \boldsymbol{\Sigma})$. The parameter of interest is ρ, the multiple correlation coefficient. Let R be the sample multiple correlation. Then it follows from the asymptotically theory of statistics (e.g., Theorem 5.2.8 of Muirhead, 1982, p. 183) that

$$\sqrt{N}(R - \rho) \sim N(0, (1 - \rho^2)^2). \qquad (6)$$

However, the distribution of R can be very skewed, and the approximation might be poor unless N is very large. Also, because $0 \le R \le 1$, it would be more effective to make a transformation for examining the fine change in R and for predicting the change. The notable Fisher's z-transformation

$$g(R) = \frac{1}{2} \log(\frac{1 + R}{1 - R})$$

is for such a purpose, and in addition, the variance of $g(R)$ does not depend on ρ. This allows us to obtain more accurate parameter estimates in modeling the behavior of R via $g(R)$. In particular, applying the delta method to equation (6) yields

$$\sqrt{N}(g(R) - g(\rho)) \sim N(0, 1). \qquad (7)$$

Also, for a given condition on (N, p, ρ) we can obtain the sample mean of $g(R)$ using Monte Carlo simulation with replications, and obtain a difference $b = \bar{g}(R) - g(\rho)$. Clearly, any bias in R will be reflected by the value of $g(R) - g(\rho)$ and consequently the value of b. We will refer b as the empirical bias of the transformed R, and develop a formula to predict the bias in R by modeling the behavior of b. A formula of corrected R will be obtained from the formula of the predicted bias.

Suppose we have obtained the values of b over N_c different sets of conditions of (N, p, ρ). They serve as the observations of the outcome variable and satisfy

$$b_i = \bar{g}(R_i) - g(\rho_i), \quad i = 1, 2, \ldots, N_c, \qquad (8)$$

where $\text{Var}(b_i) \approx \tau_i^2 = 1/(N_i N_{ri})$ with N_i and N_{ri} being the sample size and number of replications for computing b_i. For each set of conditions, we also obtained the average $\bar{R}$ across the N_r replications. We next consider building a regression model with b_i as the dependent variable and N_i, p_i and $\bar{R}_i$ as covariates, and develop formulas to predict b. We will replace $\bar{R}_i$ by R in applying the developed formulas since we do not have $\bar{R}$ in practice. Because the variance of b_i is proportional to $1/(N_i N_{ri})$, we can use the weighted least squares (WLS) method to predict the bias empirically. The WLS function is given by

$$Q_w(\boldsymbol{\beta}) = \sum_{i=1}^{N_c} w_i(b_i - \mathbf{c}_i'\boldsymbol{\beta})^2, \qquad (9)$$

where $w_i = N_i N_{ri}$, $\mathbf{c}_i$ is a vector of predictors to be identified by best-subset regression, $\boldsymbol{\beta}$ is a vector of regression coefficients to be estimated.

While the sample size N_i and the number of predictors p_i are known to affect the size of the bias, they may not have a linear relationship with the bias b_i. We consider many forms of the functions of N, p and $\bar{R}$ for predicting the value of b, and call them candidate predictors in our development. To identify the most important predictors, we consider the following eight forms of N, p and $\bar{R}$ (Yuan, Fan, & Zhao, 2019)

$$\log(x), \ x^{1/3}, \ x^{1/2}, \ x, \ x\log(x), \ x^{4/3}, \ x^{3/2}, \ x^2, \tag{10}$$

where x is N, p, or $\bar{R}$ and the transformation forms are denoted as $h_j(x)$, $j = 1$, 2, ..., 8. The eight $h_j(x)$ represent a family of power transformations on x, with power ranging from 0 to 2. In general, the bias approaches zero as the sample size increases. Consequently, we only consider potential predictors in the form of $h_j(x)/h_k(N)$, where $x = p$ and $\bar{R}$, $j, k = 1, 2, \ldots, 8$. So, there are 64 candidate predictors in the form of $h_j(p)/h_k(N)$, and 64 candidate predictors in the form of $h_j(\bar{R})/h_k(N)$. In order to use the available methods of variable selection, we mainly consider b as being predicted by a linear regression model, for example, in the form of

$$b = \beta_0 + \beta_1 \log(p)/N + \beta_2 \bar{R}^{1/2}/\log(N) + \ldots + \beta_m \bar{R}\log(\bar{R})/N^2 + \varepsilon. \tag{11}$$

Obviously, not all 128 predictors are needed for predicting the bias of the transformed R. We need to identify the most important ones out of the 128 candidates, and will use the method of best-subset regression (Haste, Tibshirani, & Friedman, 2009) for such a purpose. Our selection of predictors for calibrating b consists of the following steps. 1) Most important predictors of $b_i = \bar{g}(R_i) - g(\rho_i)$ from the 64 candidates in the form of $h_j(p_i)/h_k(N_i)$ are selected by best-subset regression, using WLS with weight $w_i = N_i N_{ri}$. Variables in each of the first 10 subsets are recorded, and each set corresponds to a model for predicting b. 2) In parallel, predictors in the form of $h_j(\bar{R}_i)/h_k(N_i)$, $j, k = 1, 2, \ldots, 8$ are selected by the method of best-subset regression, and each subset corresponds to a model for b. 3) Considering that variables in the form of $h_j(p_i)/h_k(N_i)$ may interact with those in the form of $h_j(\bar{R}_i)/h_k(N_i)$, a new subset $B(p, \bar{R})$ is formulated that includes the variables appeared in the first six[4] best subsets obtained in step 1), those in the first six best subsets obtained in step 2), and the squared term of each of these elements as well as their pairwise products. 4) Using best-subset regression, best 10 subsets are reselected from $B(p, \bar{R})$ successively, each corresponding to a different model. 5) The models identified in the previous steps are reevaluated using cross validation (CV) with independent random samples. Those that perform the best will be our new formulas for estimating ρ.

---

[4] The selected number of best subsets might be arbitrary, but our experience indicates that the additional gain becomes minimal as we select more predictors. Also, best-subset regression becomes less effective with too many variables being included in the following step that involves product terms.

Let $\hat{\boldsymbol{\beta}}$ be the WLS estimator of $\boldsymbol{\beta}$ obtained by minimizing equation (9). For a given number of predictors, the model identified by best-subset regression yields the largest R^2 defined as

$$R^2 = 1 - Q_w(\hat{\boldsymbol{\beta}})/Q_w(\hat{\beta}_0), \tag{12}$$

where $Q_w(\hat{\beta}_0)$ is the value of Q_w with only an intercept ($\mathbf{c}_i = 1$). Let $\hat{b}$ be the formula corresponding to a given set of predictors identified by the best-subset regression. The corresponding formula for estimating ρ is

$$\hat{\rho} = \frac{e^{2[g(R)-\hat{b}]} - 1}{e^{2[g(R)-\hat{b}]} + 1}, \tag{13}$$

where the element $\bar{R}$ involved in $\hat{b}$ will be replaced by the observed R. The final $\hat{\rho}$ are those that yield the smallest mean square error (MSE) or bias/variance in cross validation.

3   Conditions

Most classical results in multivariate analysis are developed based on normally distributed data. As noted earlier, we will also consider normally distributed data in this article. For the results to be valid for a wide range of conditions, we include many different conditions on N, p and ρ:

- Number of variables (p): 5, 10, 15, 20, 25, 30, 40, 50, 60, 80, 100, 120, 150, 200, a total of 14 different values.
- Multiple correlation coefficient (ρ): 0.1, 0.2, 0.3, 0.35, 0.4, 0.45, 0.5, 0.55, 0.6, 0.65, 0.7, 0.8, 0.9, 0.95, a total of 14 different values.
- Sample size (N) is nested within the condition of p: $1.6p$, $2p$, $4p$, $6p$, $8p$, $10p$, $12p$, $14p$, $16p$, $18p$, $20p$, $22p$, $25p$, $28p$, $31p$, $34p$, $37p$, $40p$, a total of 18 levels.

Thus, there are a total of $14^2 \times 18 = 3528$ different combinations of N, p and ρ. At each of the combined conditions, $N_r = 1000$ replications are used to obtain $\bar{R}$ and $\bar{g}(R) - g(\rho)$.

For a given pair of ρ and p, data in the Monte Carlo simulation are generated according to

$$y = a + b(x_1 + x_2 + \ldots + x_p) + e, \tag{14a}$$

where $a = 1$, $\mathbf{x} = (x_1, x_2, \ldots, x_p)' \sim N(\mathbf{0}, \mathbf{I})$, $e \sim N(0, \sigma^2)$. For convenience, we choose $b = 0.1$, and it follows from the population counterpart of equation (4) that

$$\rho^2 = 0.01p/[0.01p + \sigma^2] \quad \text{and} \quad \sigma^2 = 0.01p(1/\rho^2 - 1). \tag{14b}$$

Thus, we only need to choose a proper σ^2 for the simulated data to satisfy the target condition of ρ. While the process of simulating data according to (14a) and (14b) is easy to carry out, the resulting R^2 has the same property as generated by any other design that satisfies $(\mathbf{x}', y)' \sim N(\boldsymbol{\mu}, \boldsymbol{\Sigma})$ (Muirhead, 1982).

In summary, we have the values of $(b, N, p, \bar{R})$ corresponding to 3528 different conditions of (N, p, ρ). These are used to build a model with b as the dependent variable, and functions of N, p and $\bar{R}$ as the predictors.

4   Results

The results presented in this section are obtained from applying the procedures and methods described in the previous sections, with a total of 3528 values of empirical bias b. Our purpose is to identify the most important predictors for the mean bias $E[g(R)] - g(\rho)$ by conducting regression analysis on the empirical bias $b = \bar{g}(R) - g(\rho)$.

4.1   Best predictors

Following the procedure described in the method section, we first performed best-subset regression of predicting b using $h_j(p)/h_k(N)$ and $h_j(\bar{R})/h_k(N)$ according to weighted least squares, respectively. Pro Reg in SAS[5] is used to conduct the analysis. Table 1 contains the results of the best-subsets using the option of maximizing R^2 in SAS. The number of candidate predictors is listed on the left-hand side of Table 1 and the predictors identified by best-subset regression are inside the squared brackets in the main body of the table, along with the corresponding values of the R^2. To save space, only the predictors appearing in the first three subsets are reported. The results in Table 1 indicate that the single most important predictor of b in the form of $h_j(p)/h_k(N)$ is p/N, which accounts for 79.53% of the variance of b. The best-2 predictors $(p/N, p^{3/2}/N^2)$ account for 79.73% of the variance in b and the best-3 predictors $(p/N, p/N^{4/3}, p^{3/2}/N^2)$ account for 79.78% of the variance in b. Ten predictors explain 79.85% variance of b. But the size of gain in R^2 becomes minimal as more predictors are added.

Table 1. Results of Best-Subset Regression with 3528 Values of b

Covariates	Predictors, R^2
$p(64)$	$[p/N, .7953], [p/N, p^{3/2}/N^2, .7973], [p/N, p/N^{4/3}, p^{3/2}/N^2, .7978],$ ..., $[10$ predictors, $.7985]$.
$\bar{R}(64)$	$[\bar{R}^{1/3}/N^{1/2}, .2307], [\bar{R}^{1/3}/\log(N), \bar{R}^{1/2}/N^{1/2}, .3477],$ $[\log(\bar{R})/N^{4/3}, \bar{R}^{1/3}/\log(N), \bar{R}^{1/2}/N^{1/2}, .3607], ...,$ $[10$ predictors, $.4005]$.
$(p, \bar{R})(189)$	$[1$ predictor, $.8228], [2$ predictors, $.9312], [3$ predictors, $.9586],$ $[4$ predictors, $.9645], [5$ predictors, $.9684], [6$ predictors, $.9700],$ $[7$ predictors, $.9814], [8$ predictors, $.9840], [9$ predictors, $.9856],$ $[10$ predictors, $.9874]$.

The 3rd to 5th lines of Table 1 contain the information of predicting b in the form of $h_j(\bar{R})/h_k(N)$. The best single predictor is $\bar{R}^{1/3}/N^{1/2}$, which ac-

---

[5] The option "model y = v1-v64/selection = maxR stop = 10; weight w;" under Pro Reg allows us to select the best predictors from v1 to v64 according to weighted least squares.

counts for 23.07% of the variance in b. The best-10 predictors in the form of $h_j(\bar{R})/h_k(N)$ account for 40.05% of the variance of b. Thus, variables in the form of $h_j(\bar{R})/h_k(N)$ are less informative than those in the form of $h_j(p)/h_k(N)$ for predicting b.

The first six subsets in the forms of $h_j(p)/h_k(N)$ and $h_j(\bar{R})/h_k(N)$ contain 8 and 10 different predictors, respectively. These predictors are combined, together with the square of each predictor and their pairwise products. Consequently, $B(p,\bar{R})$ contains 189 ($= 18 + 18 + 18 \times 17/2$) different predictors. As described in section 2, best-subset regression is subsequently used again to select the most important predictors from $B(p,\bar{R})$. The last four lines in Table 1 report the best subsets with 1, 2, ..., 10 predictors, along with the corresponding values of R^2. The single most important predictor, $p\log(p)\log(\bar{R})/[N^{4/3}\log(N)]$, accounts for 82.28% of the variance in b and performs better than the best-10 predictors in either the form of $h_j(p)/h_k(N)$ or $h_j(\bar{R})/h_k(N)$. The set of best-2 predictors, $p\log(p)\log(\bar{R})/[N^{4/3}\log(N)]$ and $[p\log(p)]^2/[N\log(N)]^2$, accounts for 93.12% of the variance of b. The best-10 predictors out of $B(p,\bar{R})$ account for 98.74% of the variance of b. However, because both the outcome variable and the predictors involve the element $\bar{R}$, the better performance of the models identified in the last part of Table 1 may be due to capitalization on chance. We will use cross validation with independent samples to further evaluate the formulas corresponding to the identified predictors.

4.2   Validation study

We have identified 30 sets of predictors in the previous section, as listed in Table 1. In this section, their performances are further examined using independent samples. Because predictors in the form of $h_j(\bar{R})/h_k(N)$ alone do not perform well under best-subset regression, they will not be considered in the validation study. Thus, we only have 20 sets of predictors to evaluate.

Note that, for each sample, there is a predicted value of b corresponding to each set of predictors. According to equation (13), a formula for estimating the multiple correlation ρ is obtained from each of the 20 sets of predictors. For clarity in presentation, we use $\hat{\rho}_k$ to denote the estimator whose formulation has k predictors that do not involve R (i.e., only in the form of $h_j(p)/h_k(N)$), and $\tilde{\rho}_k$ to denote the estimator whose formulation has k predictors that involve $\bar{R}$ (i.e., corresponding to the last 4 lines of Table 1). In addition, we also include the original multiple correlation R and the adjusted R_{adj} in the evaluation. Thus, we have 22 formulas in total for estimating ρ. Note that $\bar{R}$ is replaced by R in the formulation of $\tilde{\rho}_k$ when evaluating its validity.

The same $N_c = 3528$ conditions on (p, ρ, N) as in the calibration study are used in the validation study. For each of the combined conditions (p, ρ, N), $N_r = 200$ replications are used to evaluate the empirical bias, variance and MSE for each of the 22 estimators. Let i be the index for the N_c conditions and j be the index for replications. The bias, variance and MSE for each estimator are computed according to

$$\text{bias}^2 = \frac{1}{N_c} \sum_{i=1}^{N_c} \text{bias}_i^2, \quad s^2 = \frac{1}{N_c} \sum_{i=1}^{N_c} s_i^2, \quad \text{MSE} = \frac{1}{N_c} \sum_{i=1}^{N_c} \text{MSE}_i, \quad (15)$$

where bias_i and s_i^2 are the empirical bias and variance across the $N_r = 200$ replications under the ith condition, and

$$\text{MSE}_i = \frac{1}{N_r} \sum_{j=1}^{N_r} (\hat{\rho}_{ij} - \rho_i)^2 = s_i^2 + \text{bias}_i^2 \quad (16)$$

with $\hat{\rho}_{ij}$ being the value of an estimator in the jth replication. In a few cases, the values of some estimators are tiny negative numbers. They are replaced by 0 in the computation.

The values of empirical mean, variance and MSE for the 22 estimators are reported in Table 2, and are ordered according to the value of MSE. Each measure is multiplied by 10^3 to include more effective digits. According to the results in the table, $\hat{\rho}_2$ yields the smallest MSE across the 3528 conditions. While R_{adj} has the smallest bias, its MSE is greater than those of eighteen of the twenty empirically corrected estimators. Among the 10 estimators whose formulations involve R, $\tilde{\rho}_3$ and $\tilde{\rho}_4$ have the smallest MSEs but the bias of $\tilde{\rho}_4$ is close to that of R_{adj}. Thus, $\tilde{\rho}_4$ is preferred over $\tilde{\rho}_3$ if one desires to have an estimator with a smaller MSE and a comparable bias to R_{adj}. The formulas of $\hat{\rho}_2$ and $\tilde{\rho}_4$ are given by equation (13) respectively via

$$\hat{b}_2 = 0.0057810 + 1.0058p/N + 1.4617p^{3/2}/N^2, \quad (17a)$$

and

$$\tilde{b}_4 = 0.011316 - 266.60 \log(p) \log(R)/N^2 + 4.0909p^{7/3} \log(p)/[N^{5/2} \log(N)]$$
$$- 43.394p \log(p) \log(R)/[N^{4/3} \log(N)] + 77.876p^{4/3} \log(R)/N^{11/6}. \quad (17b)$$

We next use graphics to examine the difference between the empirically corrected estimators $\tilde{\rho}_4$ and $\hat{\rho}_2$, and compare them against the widely used R and R_{adj}. Because the performance of these estimators are known to be affected by p, N and ρ, we will examine the changes in their empirical MSE, bias and variance as N/p and ρ vary. Note that, for the results in the validation study, there are 196 (=14 × 14) conditions of (p, ρ) while holding $N/p = r$ as a constant and 252 (=14 × 18) conditions of (p, N) while holding ρ as a constant. For each estimator, the MSEs across these 196 conditions are averaged at each value of r, and similarly for the 252 MSEs at each value of ρ. These are denoted as MSE_r and MSE_ρ, respectively. In parallel, the values of empirical variance and squared bias for each estimator are computed at each value of $r = N/p$ and ρ, and these are denoted as s_r^2, s_ρ^2, bias_r^2, and bias_ρ^2, respectively. Recall that there are 14 levels of r (1.6, 2, 4, 6, 8, 10, 12, 14, 16, 18, 20, 22, 25, 28, 31, 34, 37, 40) and 14 levels of ρ (0.1, 0.2, 0.3, 0.35, 0.4, 0.45, 0.5, 0.55, 0.6, 0.65, 0.7, 0.8, 0.9, 0.95). We will

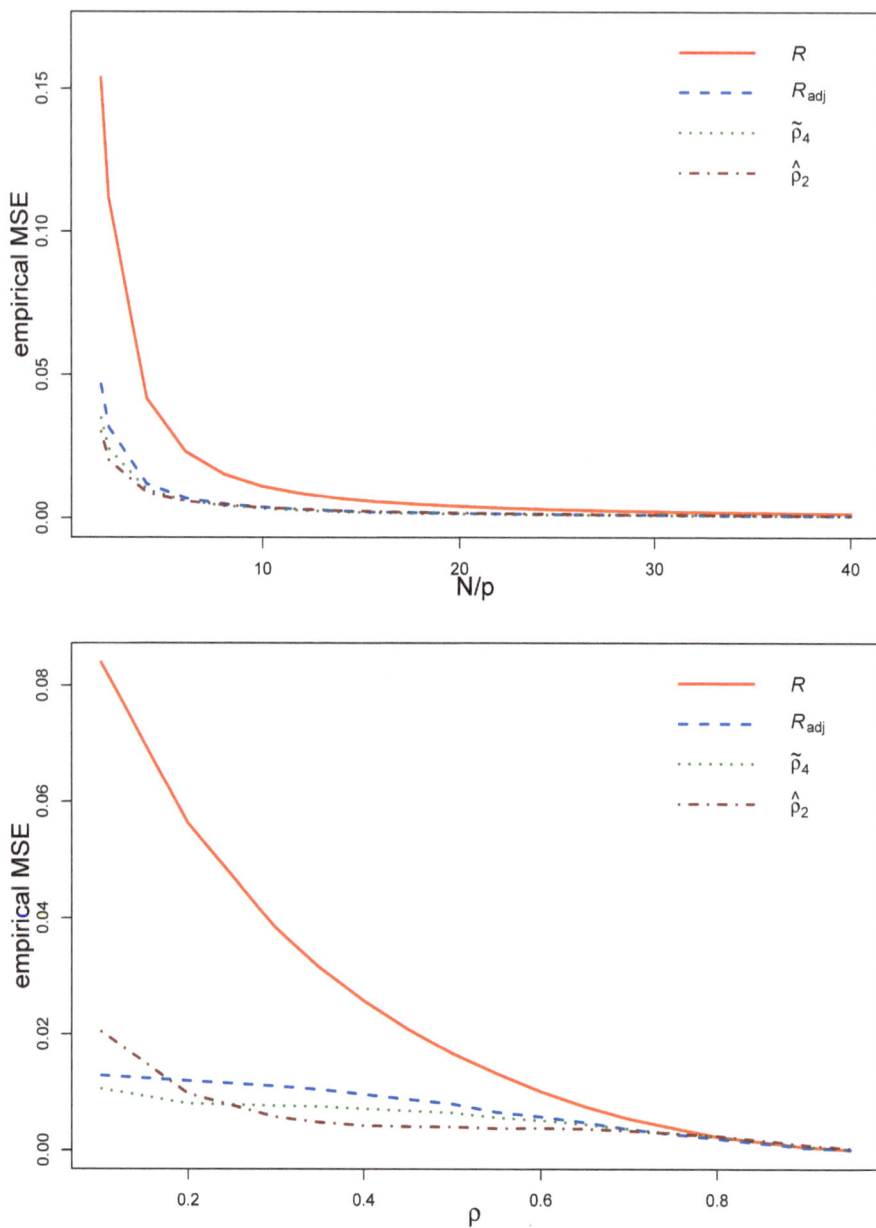

Fig. 1. Empirical MSEs of $\tilde{\rho}_4$, $\hat{\rho}_2$, R and R_{adj} against the N/p ratio (upper panel) and the population multiple correlation ρ (lower panel).

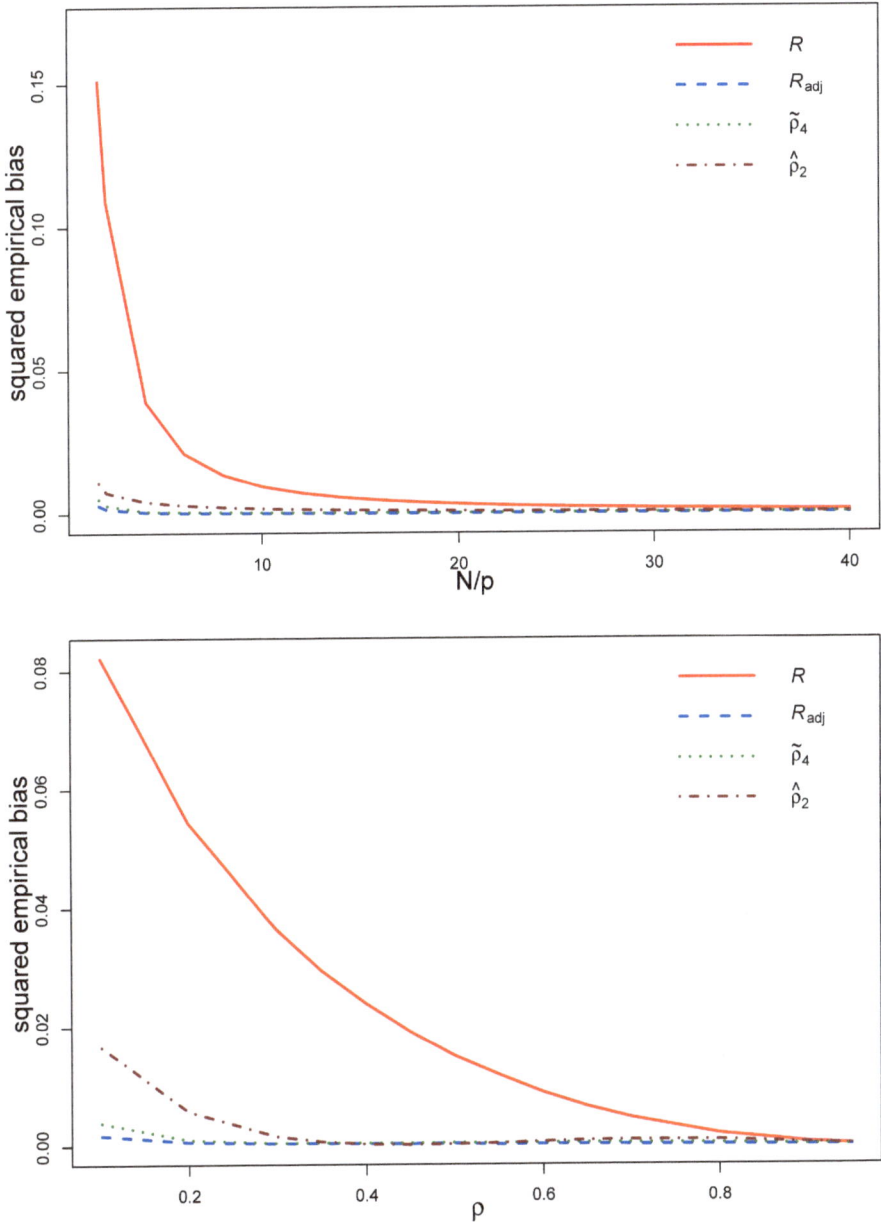

Fig. 2. Square empirical bias of $\tilde{\rho}_4$, $\hat{\rho}_2$, R and R_{adj} against the N/p ratio (upper panel) and the population multiple correlation ρ (lower panel).

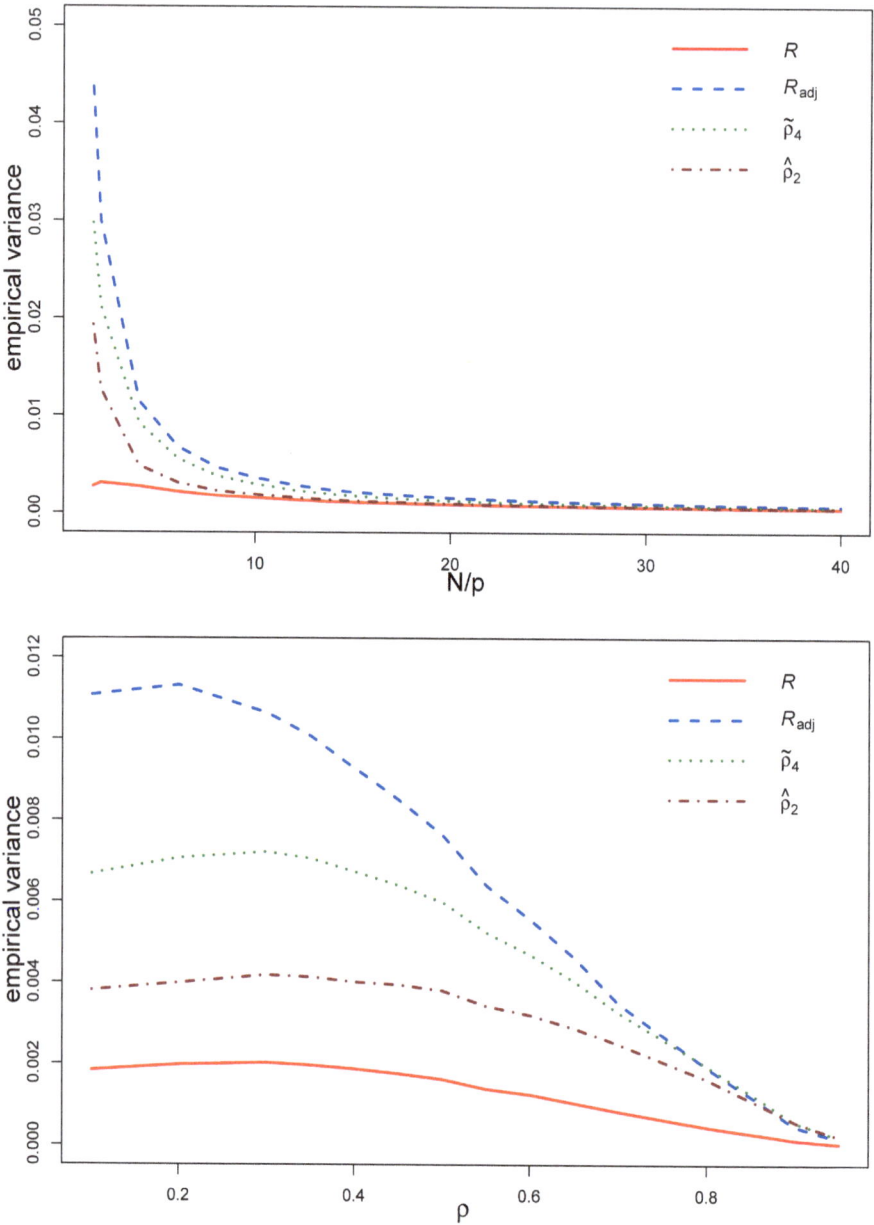

Fig. 3. Empirical variances of $\tilde{\rho}_4$, $\hat{\rho}_2$, R and R_{adj} against the N/p ratio (upper panel) and the population multiple correlation ρ (lower panel).

Table 2. MSE $\times 10^3$, $s^2 \times 10^3$, and bias$^2 \times 10^3$ of 22 Estimators across 3528 Conditions.

rank	$\hat{\rho}$	MSE	bias^2	s^2	rank	$\tilde{\rho}$	MSE	bias^2	s^2
1	$\hat{\rho}_2$	5.098	2.079	3.018	11	$\tilde{\rho}_3$	5.384	1.407	3.977
2	$\hat{\rho}_3$	5.114	2.102	3.011	12	$\tilde{\rho}_4$	5.433	0.646	4.787
3	$\hat{\rho}_6$	5.123	2.099	3.025	13	$\tilde{\rho}_5$	5.552	0.590	4.961
4	$\hat{\rho}_{10}$	5.128	2.102	3.026	14	$\tilde{\rho}_8$	5.769	0.495	5.273
5	$\hat{\rho}_9$	5.129	2.104	3.025	15	$\tilde{\rho}_9$	5.783	0.484	5.298
6	$\hat{\rho}_5$	5.132	2.102	3.030	16	$\tilde{\rho}_{10}$	5.847	0.462	5.384
7	$\hat{\rho}_4$	5.132	2.106	3.026	17	$\tilde{\rho}_6$	5.873	0.545	5.328
8	$\hat{\rho}_7$	5.133	2.104	3.029	18	$\tilde{\rho}_7$	5.891	0.513	5.379
9	$\hat{\rho}_8$	5.135	2.105	3.030	19	$\tilde{\rho}_2$	7.363	1.381	5.981
10	$\hat{\rho}_1$	5.139	2.352	2.787	20	$\tilde{\rho}_1$	9.545	2.870	6.675
	R	22.333	21.046	1.287		R_{adj}	6.847	0.341	6.506

Note: The $\hat{\rho}_k$ is an estimator whose formulation has k predictors that do not involve R, and $\tilde{\rho}_k$ is an estimator whose formulation has k predictors that involve R.

examine the fine differences among the 4 estimators by plotting their empirical values of MSE, bias^2 and s^2 against the levels of r and ρ.

Figures 1 to 3 contain the plots of the MSE, the squared bias and variance of $\tilde{\rho}_4$, $\hat{\rho}_2$, R and R_{adj} against the values of $r = N/p$ and ρ, respectively, where the red-solid line corresponds to the results of R, the blue-dashed line corresponds to those of R_{adj}, the green-dotted line corresponds to those of $\tilde{\rho}_4$, and the brown-dash-dotted line corresponds to the results of $\hat{\rho}_2$. It is clear from the three plots that R has the smallest variance but largest bias and MSE; R_{adj} has the largest or close to largest variance, smallest or close to smallest bias and a MSE not as small as $\tilde{\rho}_4$ or $\hat{\rho}_2$. Between $\hat{\rho}_2$ and $\tilde{\rho}_4$, the former has a smaller variance but larger bias. When N/p is smaller than 6, $\hat{\rho}_2$ has the smallest MSE. However, the MSE of $\tilde{\rho}_4$ becomes the smallest among the four estimators when N/p is greater than 6. As N/p increases, the differences among the four estimators become smaller, especially among the three modified ones, $\tilde{\rho}_4$, $\hat{\rho}_2$ and R_{adj}.

The plots in Figures 2 and 3 indicate that the values of the empirical bias and variance become smaller as ρ increases. This is related to the law described in equation (6) where R has little variability when ρ is close to 1. Consequently, it is easier to predict the value of R and its bias. When ρ is less than .35, $\hat{\rho}_2$ tends to be more biased than $\tilde{\rho}_4$ and R_{adj}, while $\tilde{\rho}_4$ is only slightly more biased than R_{adj}. When $\rho > .35$, all the three modified estimators become essentially unbiased.

Among the three modified estimators, $\hat{\rho}_2$ has the smallest variance. In particular, both $\hat{\rho}_2$ and $\tilde{\rho}_4$ are substantially less varied than R_{adj} when ρ is small to medium (Cohen, 1988). This implies that $\hat{\rho}_2$ and $\tilde{\rho}_4$ are more stable than R_{adj} in most cases. However, the performances of all the three modified estimators are negatively affected by small values of N/p and ρ. Although the variability of R is little affected by small values of N/p or ρ, its bias becomes more pronounced under these conditions.

Table 3 in the appendix lists the conditions of $(\rho, N, p, N/p)$ corresponding to the 35 largest MSE of $\hat{\rho}_2$ and $\tilde{\rho}_4$, R_{adj} and R, respectively. While the values of ρ of the 35 conditions for each of the three modified estimators range from .1 to .8, the values of ρ for R are only .1 and .2. The values of p for $\tilde{\rho}_4$ and R_{adj} are mostly 5 and 10 and the values of p under $\hat{\rho}_2$ and R can be up to 200, but the ratio N/p is mostly 1.6 and then 2.0. None of the 35 conditions for any of the estimators has $(N/p) > 2.0$. Thus, the ratio N/p is the key to the accuracy of each of the estimators, and a small ρ also directly causes R to behave poorly.

Table 4 contains the conditions for each of the four estimators to have the smallest MSE. The obvious feature in the table is that $\rho = .95$ for 139 out of the 140 cases. The only case is the last line under $\tilde{\rho}_4$, where $\rho = .90$. The smallest value of p under $\tilde{\rho}_4$, $\hat{\rho}_2$, R and R_{adj} are 60, 50, 40, 60, respectively. The smallest value of N/p under each of the two empirically corrected estimators is 20, and those under R_{adj} and R are 14 and 22, respectively. Thus, all the estimator perform well when ρ is close to 1.0, p is above 40 and N/p is above 14. For smaller p, these estimators can also perform well with a greater N or N/p.

5 Conclusion

In this article, under multivariate normality and using empirical modeling and statistical learning, we obtained two modified estimators of the multiple correlation ρ. These estimators perform substantially better than R and also better than R_{adj} for typical conditions in practice. The study also revealed key predictors for the bias of R with normally distributed data.

Our results suggest that $\hat{\rho}_2$ is most preferred when N/p is relatively small (e.g., less than 6), and $\tilde{\rho}_4$ is expected to work well when N/p is greater than 6. All the estimators work well when ρ is close to 1 and both N and N/p are sufficiently large.

Data in practice may not be normally distributed, and the bias and variance of R and R_{adj} will be affected by other features of the data in addition to p, $\bar{R}$ and N. In particular, it is known that the variance of R is directly related to the kurtosis of the sample (Yuan & Bentler, 2000). The behaviors of the modified estimators will also be affected by the population kurtosis, and more study is needed to better understand how the biases in these estimators are affected by the skewness and kurtosis of the underlying population or sample.

With the development of data science and the requirement of efficient and reliable analysis techniques for big data, the number of variables may exceed our imagination and also the sample size may become tremendously large. Research on bias correction for R should be extended to such conditions. Additional studies are needed in such directions.

References

Ayabe, C. R. (1985). Multicrossvalidation and the jackknife in the estimation of shrinkage of the multiple coefficient of correlation. *Educational and Psy-*

chological Measurement, *45*(3), 445–451. doi: https://doi.org/10.1177/001316448504500302

Brooks, S., & Stevens, J. (1994). Applied multivariate statistics for the social sciences. *The Statistician*, *43*(1), 219–220. doi: https://doi.org/10.2307/2348967

Browne, M. W. (1975). Predictive validity of a linear regression equation. *British Journal of Mathematical and Statistical Psychology*, *28*(1), 79–87. doi: https://doi.org/10.1111/j.2044-8317.1975.tb00550.x

Carter, D. S. (1979). Comparison of different shrinkage formulas in estimating population multiple correlation coefficients. *Educational and Psychological Measurement*, *39*(2), 261–266.

Cohen, J. (1988). *Statistical power analysis for the behavioral sciences*. New York, NY: Academic Press.

Cohen, J., & Cohen, P. (1983). *Applied multiple regression/correlation analysis for the behavioral sciences*. Hillsdale, NJ: Lawrence Erlbaum.

Cummings, C. C. (1982). *Estimates of multiple correlation coefficient shrinkage*. Paper presented at the annual meeting of the American Educational Research Association, New York. (ERIC Document Reproduction Service No. ED 220 517).

Darlington, R. B. (1968). Multiple regression in psychological research and practice. *Psychological Bulletin*, *69*(3), 161-182. doi: https://doi.org/10.1037/h0025471

Ezekiel, M. (1929). The application of the theory of error to multiple and curvilinear correlation. *Journal of the American Statistical Association*, *24*(165A), 99–104. doi: https://doi.org/10.1080/01621459.1929.10506278

Haste, T., Tibshirani, R., & Friedman, J. (2009). *The elements of statistical learning (2nd)*. Springer.

Huberty, C. J., & Mourad, S. A. (1980). Estimation in multiple correlation/prediction. *Educational and Psychological Measurement*, *40*(1), 101–112. doi: https://doi.org/10.1177/001316448004000113

Kromrey, J. D., & Hines, C. V. (1995). Use of empirical estimates of shrinkage in multiple regression: a caution. *Educational and Psychological Measurement*, *55*(6), 901–925. doi: https://doi.org/10.1177/0013164495055006001

Krus, D. J., & Fuller, E. A. (1982). Computer assisted multicrossvalidation in regression analysis. *Educational and Psychological Measurement*, *42*(1), 187–193. doi: https://doi.org/10.1177/0013164482421019

Kvålseth, T. O. (1985). Cautionary note about R^2. *The American Statistician*, *39*(4), 279–285. doi: https://doi.org/10.2307/2683704

Larson, S. C. (1931). The shrinkage of the coefficient of multiple correlation. *Journal of Educational Psychology*, *22*(1), 45-55. doi: https://doi.org/10.1037/h0072400

Lord, F. M. (1950). Efficiency of prediction when a regression equation from one sample is used in a new sample. *ETS Research Bulletin Series*, *1950*(2), i–6. doi: https://doi.org/10.1002/j.2333-8504.1950.tb00478.x

Muirhead, R. J. (1982). *Aspects of multivariate statistical theory.* Hohokcn, New Jersey: Wiley.

Nicholson, G. E. (1948). *The application of a regression equation to a new sample* (Unpublished doctoral dissertation). University of North Carolina at Chapel Hill.

Stein, C. (1960). Multiple regression contributions to probability and statistics. *Essays in Honor of Harold Hotelling, 103.*

Wherry, R. (1931). A new formula for predicting the shrinkage of the coefficient of multiple correlation. *The Annals of Mathematical Statistics, 2*(4), 440–457. doi: https://doi.org/10.1214/aoms/1177732951

Yin, P., & Fan, X. T. (2001). Estimating R^2 shrinkage in multiple regression: a comparison of different analytical methods. *The Journal of Experimental Education, 69*(2), 203–224. doi: https://doi.org/10.1080/00220970109600656

Yuan, K.-H., & Bentler, P. (2000). Inferences on correlation coefficients in some classes of nonnormal distributions. *Journal of Multivariate Analysis, 72,* 230-248. doi: https://doi.org/10.1006/jmva.1999.1858

Yuan, K.-H., Fan, C., & Zhao, Y. (2019). What causes the mean bias of the likelihood ratio statistic with many variables? *Multivariate Behavioral Research, 54*(6), 840–855. doi: https://doi.org/10.1080/00273171.2019.1596060

Yuan, K.-H., Tian, Y., & Yanagihara, H. (2015). Empirical correction to the likelihood ratio statistic for structural equation modeling with many variables. *Psychometrika, 80*(2), 379-405. doi: https://doi.org/10.1007/s11336-013-9386-5

Appendix

The two tables in this appendix contain 35 conditions for $\tilde{\rho}_4$, $\hat{\rho}_2$, R and R_{adj} to have the largest and smallest MSE, respectively. These are discussed at the end of section 4.

Table 3. Conditions Corresponding to the 35 (1%=35/3528) Largest Values of MSE ($\times 10^2$) of $\tilde{\rho}_4$, $\hat{\rho}_2$, R and R_{adj}.

#	$\tilde{\rho}_4$					$\hat{\rho}_2$					R					R_{adj}				
	MSE	ρ	N	p	N/p	MSE	ρ	N	p	N/p	MSE	ρ	N	p	N/p	MSE	ρ	N	p	N/p
1	21.26	.10	8	5	1.6	22.93	.10	8	5	1.6	57.76	.10	8	5	1.6	25.67	.10	8	5	1.6
2	15.81	.30	8	5	1.6	15.42	.20	8	5	1.6	51.07	.10	16	10	1.6	18.65	.20	8	5	1.6
3	15.36	.20	8	5	1.6	14.10	.30	8	5	1.6	50.30	.10	24	15	1.6	18.25	.30	8	5	1.6
4	15.35	.65	8	5	1.6	12.67	.35	8	5	1.6	50.03	.10	40	25	1.6	17.13	.35	8	5	1.6
5	14.75	.70	8	5	1.6	11.41	.10	16	10	1.6	49.59	.10	48	30	1.6	15.72	.40	8	5	1.6
6	14.73	.35	8	5	1.6	11.38	.10	10	5	2.0	48.95	.10	32	20	1.6	14.72	.50	8	5	1.6
7	14.54	.60	8	5	1.6	11.31	.40	8	5	1.6	48.87	.10	80	50	1.6	14.69	.45	8	5	1.6
8	14.31	.50	8	5	1.6	10.80	.10	24	15	1.6	48.78	.10	64	40	1.6	14.56	.60	8	5	1.6
9	14.03	.55	8	5	1.6	10.48	.45	8	5	1.6	48.52	.10	160	100	1.6	14.28	.10	10	5	2.0
10	13.73	.40	8	5	1.6	10.43	.50	8	5	1.6	48.38	.10	96	60	1.6	14.27	.65	8	5	1.6
11	13.29	.45	8	5	1.6	10.17	.60	8	5	1.6	48.30	.10	128	80	1.6	14.04	.55	8	5	1.6
12	12.94	.80	8	5	1.6	9.69	.70	8	5	1.6	48.21	.10	240	150	1.6	13.44	.70	8	5	1.6
13	12.51	.50	10	5	2.0	9.62	.65	8	5	1.6	48.20	.10	192	120	1.6	13.12	.10	16	10	1.6
14	12.01	.45	10	5	2.0	9.45	.10	40	25	1.6	48.11	.10	320	200	1.6	13.06	.20	10	5	2.0
15	11.97	.10	10	5	2.0	9.29	.55	8	5	1.6	42.19	.10	10	5	2.0	12.95	.50	10	5	2.0
16	11.76	.70	10	5	2.0	9.27	.20	10	5	2.0	41.61	.20	8	5	1.6	12.42	.45	10	5	2.0
17	11.39	.55	10	5	2.0	8.99	.10	48	30	1.6	39.86	.10	20	10	2.0	12.40	.30	10	5	2.0
18	11.32	.60	10	5	2.0	8.81	.20	16	10	1.6	39.09	.20	16	10	1.6	12.20	.45	16	10	1.6
19	11.24	.35	10	5	2.0	8.73	.10	20	10	2.0	39.08	.10	30	15	2.0	12.18	.35	10	5	2.0
20	10.99	.30	10	5	2.0	8.51	.10	320	200	1.6	38.21	.10	80	40	2.0	12.14	.10	24	15	1.6

Table 3. Conditions Corresponding to the 35 (1%=35/3528) Largest Values of MSE ($\times 10^2$) of $\tilde{\rho}_4$, $\hat{\rho}_2$, R and R_{adj} (Continued).

#	$\tilde{\rho}_4$					$\hat{\rho}_2$					R					R_{adj}				
	MSE	ρ	N	p	N/p	MSE	ρ	N	p	N/p	MSE	ρ	N	p	N/p	MSE	ρ	N	p	N/p
21	10.99	.10	20	10	2.0	8.44	.10	160	100	1.6	38.19	.20	32	20	1.6	11.85	.20	16	10	1.6
22	10.96	.20	10	5	2.0	8.37	.10	32	20	1.6	38.10	.10	100	50	2.0	11.47	.55	10	5	2.0
23	10.72	.65	10	5	2.0	8.37	.10	240	150	1.6	38.06	.10	40	20	2.0	11.38	.40	10	5	2.0
24	10.48	.40	10	5	2.0	8.35	.80	8	5	1.6	37.77	.10	200	100	2.0	11.34	.50	16	10	1.6
25	10.44	.10	16	10	1.6	8.27	.10	192	120	1.6	37.65	.10	400	200	2.0	11.32	.60	10	5	2.0
26	10.01	.10	24	15	1.6	8.24	.10	96	60	1.6	37.63	.10	160	80	2.0	11.20	.80	8	5	1.6
27	8.94	.45	16	10	1.6	8.21	.70	10	5	2.0	37.61	.10	120	60	2.0	11.09	.10	20	10	2.0
28	8.81	.20	16	10	1.6	8.20	.30	10	5	2.0	37.54	.10	50	25	2.0	11.09	.70	10	5	2.0
29	8.25	.50	16	10	1.6	8.19	.10	64	40	1.6	37.48	.10	60	30	2.0	11.04	.30	16	10	1.6
30	7.98	.55	16	10	1.6	8.18	.10	80	50	1.6	37.37	.10	240	120	2.0	10.74	.35	16	10	1.6
31	7.88	.65	16	10	1.6	8.11	.10	128	80	1.6	37.14	.10	300	150	2.0	10.62	.40	16	10	1.6
32	7.87	.30	16	10	1.6	8.08	.45	10	5	2.0	37.10	.20	48	30	1.6	10.13	.55	16	10	1.6
33	7.82	.10	30	15	2.0	7.66	.50	10	5	2.0	37.03	.20	80	50	1.6	10.10	.65	16	10	1.6
34	7.63	.10	40	25	1.6	7.49	.35	10	5	2.0	36.85	.20	40	25	1.6	9.95	.35	24	15	1.6
35	7.55	.60	16	10	1.6	7.46	.65	10	5	2.0	36.82	.20	24	15	1.6	9.84	.65	10	5	2.0

Table 4. Conditions Corresponding to the 35 ($1\% = 35/3528$) Smallest Values of MSE ($\times 10^2$) of $\tilde{\rho}_4$, $\hat{\rho}_2$, R and R_{adj}.

#	$\tilde{\rho}_4$					$\hat{\rho}_2$					R					R_{adj}				
	MSE	ρ	N	p	N/p	MSE	ρ	N	p	N/p	MSE	ρ	N	p	N/p	MSE	ρ	N	p	N/p
1	1.14	.95	8000	200	40	4.27	.95	8000	200	40	2.73	.95	8000	200	40	1.14	.95	8000	200	40
2	1.35	.95	7400	200	37	4.45	.95	6000	150	40	3.23	.95	6000	150	40	1.31	.95	7400	200	37
3	1.52	.95	6000	150	40	4.54	.95	7400	200	37	3.27	.95	4800	120	40	1.51	.95	6000	150	40
4	1.53	.95	5550	150	37	5.01	.95	5550	150	37	3.36	.95	7400	200	37	1.51	.95	5550	150	37
5	1.57	.95	6800	200	34	5.58	.95	6800	200	34	3.36	.95	5550	150	37	1.53	.95	6800	200	34
6	1.65	.95	6200	200	31	5.63	.95	4800	120	40	3.40	.95	4000	100	40	1.53	.95	6200	200	31
7	2.00	.95	4800	120	40	5.79	.95	3700	100	37	3.64	.95	6800	200	34	1.80	.95	4400	200	22
8	2.15	.95	4000	100	40	5.79	.95	4000	100	40	3.97	.95	3700	100	37	1.84	.95	5000	200	25
9	2.20	.95	3700	100	37	5.84	.95	4440	120	37	4.04	.95	4440	120	37	2.01	.95	4800	120	40
10	2.25	.95	5100	150	34	5.85	.95	2400	60	40	4.14	.95	6200	200	31	2.03	.95	5600	200	28
11	2.25	.95	5600	200	28	6.08	.95	6200	200	31	4.17	.95	5100	150	34	2.11	.95	4000	100	40
12	2.26	.95	4440	120	37	6.25	.95	3400	100	34	4.60	.95	3200	80	40	2.15	.95	4000	100	40
13	2.38	.95	3720	120	31	6.29	.95	3200	80	40	4.62	.95	2960	80	37	2.19	.95	3700	100	37
14	2.42	.95	4200	150	28	6.56	.95	5100	150	34	4.84	.95	3720	120	31	2.24	.95	5100	150	34
15	2.57	.95	5000	200	25	6.82	.95	4080	120	34	4.87	.95	4200	150	28	2.25	.95	4440	120	37
16	2.63	.95	4650	150	31	6.89	.95	3720	120	31	4.96	.95	4650	150	31	2.28	.95	3720	120	31
17	2.75	.95	3400	100	34	7.19	.95	5000	200	25	5.10	.95	3400	100	34	2.39	.95	4200	150	28
18	2.76	.95	4400	200	22	7.19	.95	2960	80	37	5.14	.95	4080	120	34	2.41	.95	3300	150	22
19	2.81	.95	3360	120	28	7.39	.95	4650	150	31	5.14	.95	5600	200	28	2.56	.95	4650	150	31
20	2.98	.95	4080	120	34	7.44	.95	5600	200	28	5.40	.95	2800	100	28	2.65	.95	3400	100	34

Table 4. Conditions Corresponding to the 35 (1%=35/3528) Smallest Values of MSE ($\times 10^2$) of $\tilde{\rho}_4$, $\hat{\rho}_2$, R and R_{adj} (Continued).

#	$\tilde{\rho}_4$					$\hat{\rho}_2$					R					R_{adj}				
	MSE	ρ	N	p	N/p	MSE	ρ	N	p	N/p	MSE	ρ	N	p	N/p	MSE	ρ	N	p	N/p
21	3.08	.95	3200	80	40	7.80	.95	3100	100	31	5.43	.95	3360	120	28	2.66	.95	3750	150	25
22	3.13	.95	2960	80	37	8.22	.95	2480	80	31	5.56	.95	2720	80	34	2.71	.95	3360	120	28
23	3.19	.95	3300	150	22	8.35	.95	2220	60	37	5.62	.95	2220	60	37	2.92	.95	4080	120	34
24	3.25	.95	3750	150	25	8.35	.95	2000	50	40	5.65	.95	2400	60	40	3.07	.95	3200	80	40
25	3.36	.95	2800	100	28	8.46	.95	3750	150	25	5.95	.95	2000	50	40	3.14	.95	3600	200	18
26	3.54	.95	2400	60	40	8.66	.95	3360	120	28	6.11	.95	3100	100	31	3.15	.95	2960	80	37
27	3.54	.95	3100	100	31	8.67	.95	4200	150	28	6.17	.95	1850	50	37	3.33	.95	3200	200	16
28	3.63	.95	4000	200	20	8.73	.95	2720	80	34	6.31	.95	1600	40	40	3.34	.95	3000	120	25
29	3.75	.95	3000	120	25	9.10	.95	2040	60	34	6.55	.95	2480	80	31	3.39	.95	3100	100	31
30	3.93	.95	2720	80	34	9.10	.95	4400	200	22	6.59	.95	2040	60	34	3.46	.95	3000	150	20
31	3.97	.95	2480	80	31	9.22	.95	1850	50	37	6.59	.95	5000	200	25	3.49	.95	2800	100	28
32	4.20	.95	2220	60	37	9.78	.95	3000	120	25	7.06	.95	3750	150	25	3.51	.95	2400	60	40
33	4.26	.95	2640	120	22	9.97	.95	1700	50	34	7.25	.95	3000	120	25	3.62	.95	2700	150	18
34	4.49	.95	2500	100	25	10.22	.95	3300	150	22	7.31	.95	4400	200	22	3.76	.95	2800	200	14
35	4.72	.90	8000	200	40	10.23	.95	4000	200	20	7.51	.95	2240	80	28	3.82	.95	2480	80	31

www.ingramcontent.com/pod-product-compliance
Lightning Source LLC
Chambersburg PA
CBHW041313210326
41599CB00008B/258